GLOSSARY

Standardized Terminology for Doll Collectors

United Federation of Doll Clubs, Inc.
1978

COVER:

Parian-type doll with glass eyes and decorated porcelain yoke. The body is cloth with porcelain Kintzback type hands. This doll was shown in *More about Dolls*, by Janet Johl published in 1946 when the doll belonged to Mary E. Lewis. In 1947, Mary E. Lewis gave first place to this doll in her book, *The Marriage of Diamonds and Dolls.* The provenance of this doll as given by Mary E. Lewis is questionable. The head was probably made in 1870s at the same time as the hands. It is not known where it was made.

ISBN 0–87588–142–4

Printed in the United States of America

The Glossary Committee

The Glossary Committee was appointed by the President of the United Federation of Doll Clubs (U.F.D.C.), Mrs. Stephen Orth, in 1975. It continued to function under the aegis of the current President, Dr. Candace Doelman. Thanks to Dr. Doelman and the membership of U.F.D.C. the Glossary Committe was able to prepare this booklet to standardize terminology for all collectors of dolls.

The Committee members are as follows:

Larry L. Belles,
Chief Curator of Collections, Margaret Woodbury Strong Museum
Co-chairman of the Museum Committee, U.F.D.C.

Dorothy S. Coleman,
Chairman of the Glossary Committee, U.F.D.C.
Chairman of the Research Committee, U.F.D.C.

Elizabeth Ann Coleman,
Curator, Brooklyn Museum
Chairman of the Speakers' Bureau, U.F.D.C.

Sarah Kocher,
Co-chairman of Exhibits, U.F.D.C.

Z. Frances Walker,
Co-chairman of Exhibits, U.F.D.C.

Margaret Whitton,
Curator, Margaret Woodbury Strong Museum
Chairman of Judges, U.F.D.C.
Co-chairman of Museum Committee, U.F.D.C.

The function of the Glossary Committee has been to standardize terminology in accordance with the preferences of the majority of doll collectors. In order to accomplish this it was necessary to prepare a questionnaire with choices for controversial terms. A sample of knowledgeable collectors representing various regions and doll related occupations was chosen. All U.F.D.C. members as well as the nearly ten thousand subscribers to the *Doll Reader* were also invited to participate. The nearly three hundred questionnaires that were completed and returned to the Committee were tabulated. Each term that received a majority of the votes had to be defined in detail by the Glossary Committee. These definitions are put into the Glossary booklet so that all collectors can have ready access to the terms that were chosen by the collectors themselves.

It is advisable to read the Foreword for a better understanding of this Glossary. Rejected terms can usually be found in the index if they are not one of the entries.

Acknowledgements

The entries in the Glossary are in accordance with the results of a questionnaire survey in which all collectors in the United Federation of Doll Clubs were invited to participate. An invitation was also extended to the subscribers of the *Doll Reader*. The Glossary Committee wishes to express its gratitude for the thoughtful answers and opinions expressed by the respondees to the questionnaire. They gave generously of their time and knowledge so that all collectors of dolls might enjoy the benefits recorded in this booklet. The terms selected by the participants should provide better understanding between and among collectors and greater enjoyment of the hobby.

The pictures illustrating the terms are all photographs of dolls at the Margaret Woodbury Strong Museum, Rochester, New York. These pictures were taken by Barbara Jendricks of the Margaret Woodbury Strong Museum staff. The Glossary Committee is most appreciative of the privileges exteded to it by the Margaret Woodbury Strong Museum.

Thanks should also go to Gary Ruddell for technical assistance.

Foreword

As early as 1965 the United Federation of Doll Clubs realized that there was a need for a Glossary Committee to standardize terminology as applied to dolls. Through the years terms and names evolved in a haphazard and consequently confusing manner. The terms often differed markedly from one area to another and even from person to person within areas. For example, "Godey" is a term applied by collectors to half a dozen different types of molded hairdos on dolls. On the other hand, several different names are applied to the same type of doll. For example a Frozen Charlotte, a bathing baby, and a pillar doll are all names used by collectors for the same type of doll.

Problems concerning the names of dolls were recognized over twenty years ago by Janet Johl who wrote in *Doll News*, March 1955, page 22:

"If a dealer or collector cannot produce the authentic data to substantiate calling a doll a certain type, then it is surely best to leave the doll unnamed... This is the type of erroneous data we must find and bring to light, it is comparable to stating that a doll is a Belton because it has two holes in the head."

Authors are especially responsible as they often coin names for a particular type of doll. Some of these names have no pertinent meaning but they frequently become widely used because they have appeared in print. An excellent example of this is the term "Milliners' Model." This term is applied to dolls that have no relation whatsoever to either milliners or models. When these so-called "Milliners' Model" dolls were originally made, the term "Milliner" was applied primarily to a dressmaker or modiste. However, records indicate that these so-called "Milliners' Model" dolls were used only as playthings for children and were never used by milliners or modistes to display fashionable attire. The term "Milliners' Model" was originated by Eleanor St. George in the early 1950s, when it appears in her books which were published at least a century after this type of doll was made. It is not surprising, but it is indicative of the thoughtful study by the doll collectors responding to the questionnaire survey, that most respondees voted against the continued use of such a misleading term.

There are numerous precedents for changes in doll terminology over the years. This is good because it results from increased knowledge about dolls. In the 1930s many doll collectors referred to pink china dolls' heads as "Staffordshire" and fine quality white china dolls' heads as "Chelsea." All of these dolls' heads supposedly originated in England. However we now know that most of our china dolls and dolls' heads came from Germany and not a single glazed china headed doll has been proven by collectors of dolls to have been made in England. Thus the terms "Staffordshire" and "Chelsea" have become obsolete and have ceased to be used.

In 1975 the Glossary Committe was reactivated with the preparation of a survey questionnaire designed to explore the necessity and feasibility of standardizing terminology relating to dolls. During the years 1965-75 the ever-increasing number of identifying and/or descriptive terms as applied to dolls had become an even more burdensome task to comprehend. Both the number of collectors of dolls and the amount of printed publications pertaining to dolls had grown to great numbers; thus increasing the need for standardizing terminology.

A lengthy questionnaire was designed to cover a variety of topics and especially the most controversial ones. Selections and definitions of terms in the Glossary are based on the questionnaire returns. Professional sampling methods were used in order to obtain as representative a group of opinions as possible. The sample included doll collectors, authors, museum curators, dealers and officials of the United Federation of Doll Clubs. All geographic areas of the United States were included and there were even small groups in Great Britain and Canada. The samples from each of these groups were studied separately to make sure that there were no significant differences in the selections geographically or occupationally. It is hoped that the representative sample thus obtained reflects the general opinion of doll collectors. In order to make the study as comprehensive as possible invitations to participate were published in both *Doll News* and the *Doll Reader.* Thus all members of the United Federation of Doll Clubs and many other collectors received an invitation to participate. The Committee wished that every interested person who so desired could have the opportunity of expressing themselves on the questionnaire. The additional respondees who accepted the invitation followed the same patterns as the earlier survey showed and served to reinforce the selections.

The function of the Committee did not include the examination of all terms that are applied to dolls. But the Committee added a few non-controversial terms such as "Snow Babies," the exact meaning of which might not be known to a new collector. The Committee tried to write clear and concise definitions for each of the chosen terms. In many cases explanations were written for the rejected terms. Terms are listed in the Glossary alphabetically. If the term you seek cannot be found in the Glossary, check the index. This will often show you the rejected term and the entry page of the preferred term that took its place. In order to augment the clarity of the terms, the Committee chose pictures that would best illustrate most of the definitions. All of the many pictures were taken by Barbara Jendricks at the Margaret Woodbury Strong Museum.

The survey participants generally voted in favor of the English version of a name rather than the French, German or other variant, but this was not always true as for example creche was chosen over crib. Trade names are the most accurate terms but some of these were rejected. "Peg Wooden" as a descriptive term for a type of doll received more votes than "Dutch" dolls, although in times past these dolls were always referred to as "Dutch Dolls." The survey showed overwhelmingly that people do **not** want dolls called "French" if it cannot be proven that they were made entirely in France, regardless of the fine quality of the doll. Since most dolls are international in their origins, this mandate presents some problems. Many dolls with unknown origins have been termed "French" in the past.

The majority vote determined the selection of each term. Many of these terms you as an individual or even the members of the Glossary Committee would not have chosen without this majority mandate from hundreds of collectors who participated in the survey.

You may not agree with all of the selections and wish to continue using your familiar names, but you should be aware of and know the standardized terms as recognized by the United Federation of Doll Clubs. These terms should be used in dealing in dolls, exhibiting dolls, writing about them or discussing them in public, so that others can understand you clearly and accurately. The unfounded attribution of unmarked dolls to a well-known or prestigious maker is a practice that should be discontinued.

The United Federation of Doll Clubs wishes to maintain only the highest standards of integrity in the field of doll collecting. It is hoped that this Glossary will enable all those interested in the study and collecting of dolls to be more accurate and precise in their identifying terminology.

The Glossary Committe wishes to express its thanks to the nearly three hundred people who supported this project by providing their preferences and thoughtful opinions. The Glossary Committee is pleased to present this report to the members and Board of Management of the United Federation of Doll Clubs and sincerely hopes that the contents of this publication will further unify doll collectors in their mutual pursuits.

The Glossary Committee

L.L. Belles
D.S. Coleman
E.A. Coleman
S. Kocher
Z.F. Walker
M. Whitton

Glossary

NOTE: Terms that are in italics can be found listed as separate entries in this glossary.

A. M.: The abbreviation for Armand Marseille. Collectors sometimes use A. & M. which is not correct. Illustration 1.

1. A. M. mark stands for Armand Marseille. Mold 670, bisque shoulder head.

ACTION KEWPIES: Not selected to be included as dolls by collectors who participated in the questionnaire survey. These Kewpies are not articulated and are really figurines. Illustration 2.

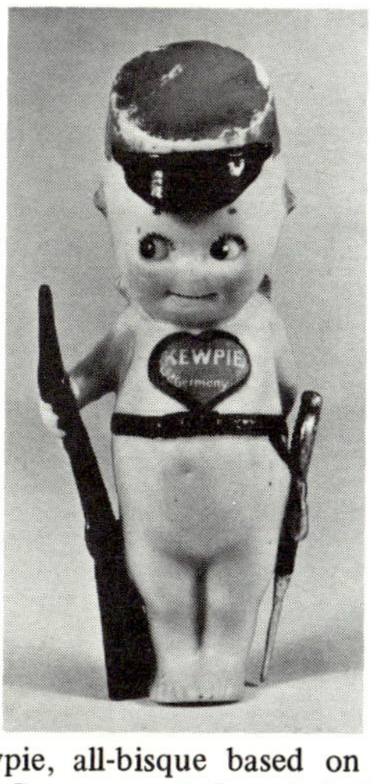

2. Action Kewpie, all-bisque based on Rose O'Neill's design, made in Germany, © Joseph L. Kallus.

AGE OF DOLLS: Name and time period selected by collectors:

Antique, over 75 years
Collectible, 25-75 years
Modern, under 25 years

The United States Customs uses 100 years or more in their definition of Antiques, the Canadian Customs uses 50 years or more. The time spans listed above are rough approximations since it is very difficult to estimate the precise age of a doll. Also see *Dating a Doll*.

AMERICAN DOLLS: Dolls made entirely in America. Illustration 3.

3. American doll, all-composition, marked on back "Amberg L. A. & S. © 1928."

AMERICAN-TYPE DOLLS: Dolls designed in America, part or all of the dolls made elsewhere, such as bisque Kewpies, Bye-Los, and so forth.
Illustrations **4A** and **4B**.

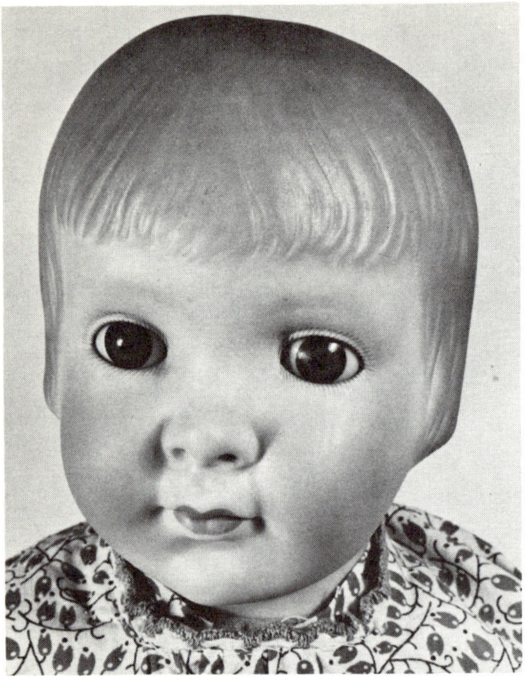

4A. American-type doll, bisque head designed in America by Grace C. Rockwell, made in Germany.

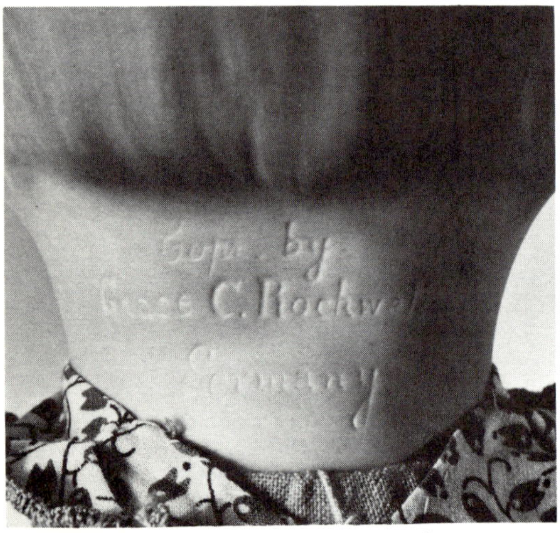

4B. American-type doll designed by an American, head made in Germany according to this mark, "Copr. by Grace C. Rockwell, Germany."

ANIMAL-FACE TOYS: Not classified as dolls. This includes the two-faced variations with one human face and one animal face on an animal-type body. Also see *Doll.* Illustration **5**.

5. Animal-Face Toys, all-composition, pig made by Madame Alexander, cat unmarked.

ANTIQUE DOLLS: Dolls made approximately 75 years or more ago. For methods of dating dolls see *Dating a Doll.*

APOLLO KNOT: Elaborate style of hair-do most often found on the *Early Molded-Hair Papier-Mâché* head dolls of the 1830s. This hair style copies contemporary arrangements which frequently required wire loop supports. Illustration **6**.

6. Apollo Knot hair style, early molded hair papier-mâché head, kid body, wooden limbs.

ARTICULATION: The ability of a doll to move in one or more joints. It does not pertain to the ability of the doll to talk.

AUTHENTIC: Of undisputed origin.

AUTHENTIC CLOTHES: As used by collectors, are clothes that are not necessarily original but are in a style contemporary with the period of the doll. Not only the style but also the fabrics and trims are suitable for the period represented.

AUTHENTICATE: In order to authenticate a doll it is usually necessary to verify and interpret the marks on the doll and to make sure the doll is not a reproduction. Marks are often difficult to read and even so-called "Experts" can sometimes make mistakes in deciphering a mark. Also see *Marks.*

BALD HEADS: Heads of china or any other material which are dome shaped and have no holes cut in the pate. These heads usually have a wig. Illustration **7**.

7. Bald heads, bisque head on left, china head on right.

BALL-JOINTED COMPOSITION BODY: Composition body with ball joints for the articulation. The body is usually shaped to resemble that of a young child. The *Trade Name* for dolls having these bodies was "Bébé incassable" or "Unbreakable child (or baby)"; but collectors well know that the bodies can be broken. In the 20th century this type of body was adapted for dolls representing older children and adults. Illustration **8**.

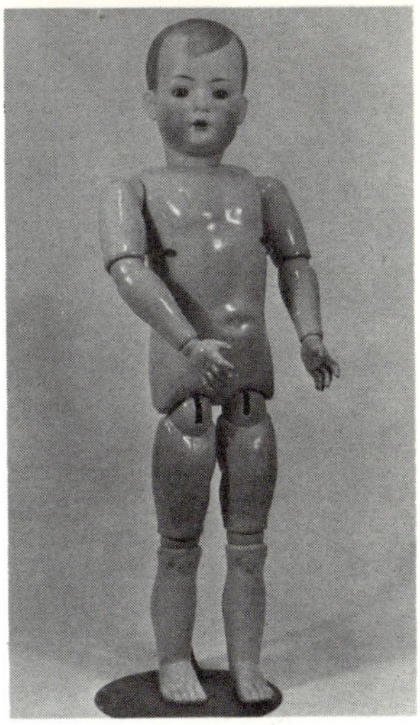

8. Ball-jointed composition body, bisque head marked "B S W" in a heart for Bruno Schmidt of Waltershausen.

BELTON-TYPE HEADS: Bisque heads made with one, two or three small holes in the pate, but otherwise uncut. The holes were probably used for stringing. No connection with any person named "Belton" has been found. Evidence suggests that this type of head was widely produced in the late 19th century, after 1875. Illustration **9**.

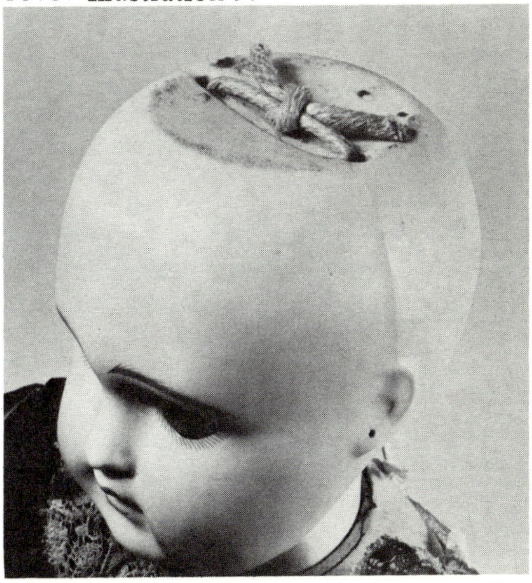

9. Belton-type head, top of bisque head is concave. Top of head could be convex or flat.

BENT-LIMB BODY: Jointed composition body made with five pieces and having all four limbs bent. The body is shaped like that of a baby about six to nine months old. This type of body first appeared around 1909. For example the K✡R mold 100 Baby. Illustration **10**.

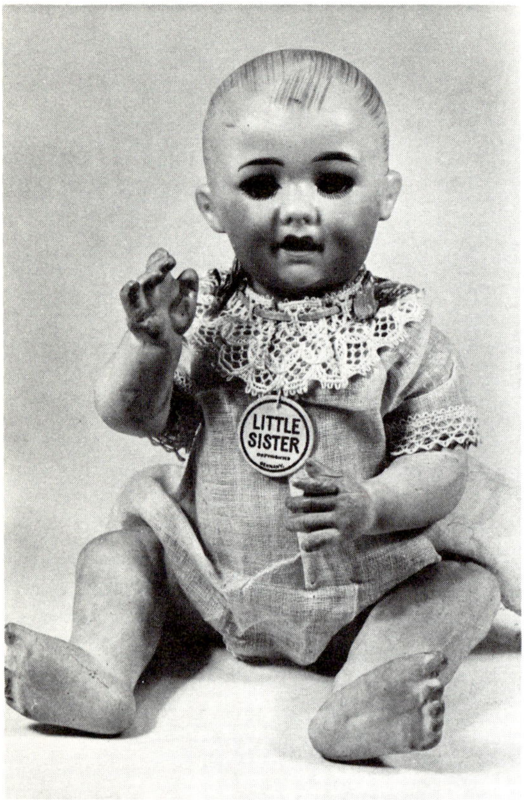

10. Bent-limb body of composition, bisque head marked "Germany G326 B A 1 M DRGM."

BIEDERMEIER HEADS: Term not chosen by collectors, see *Bald Heads*.

BISQUE DOLLS: Unglazed *Porcelain* dolls, usually tinted. These dolls have a matte surface finish. The term applies to bisque-head dolls as well as all-bisque dolls.

BONE DOLLS: Dolls with bone (ivory) heads or made completely of bone (ivory). Illustration **11**.

BONNET DOLLS: See *Hatted Dolls*.

11. Bone or ivory doll, probably eighteenth century.

BOUDOIR DOLLS: Long-limbed dolls, usually with cloth bodies and made in the 1920s. The heads are cloth or composition, often a mask-type face. The costumes are generally either Flapper or mid-19th century period styles or clowns. The *Trade Names* were "Boudoir Doll," "Flapper Doll" or "French Doll," but this did not mean they were made in France. Illustration **12**.

12. Boudoir doll, Lenci with felt face and long limbs, original period costume.

BROTEIG: German term for a composition made of rye flour, glue and water. This was used for making dolls and should simply be called "composition."

CAN'T BREAK 'EM: Dolls with composition heads, see *Cold-Press Composition.*

CARTON MOULE: French term not selected by collectors, see *Molded Cardboard.*

CASE DOLL: Dressed doll(s), permanently housed in a case with a glass front and usually glass on the sides. Often there are flowers or other decorations in the case. Some of these case dolls are found in churches as thank offerings, but the majority were created for decoration in the home. Illustration **13**.

13. Case Doll, wax-over-composition head; probably original clothes and case.

CELLULOID DOLLS: Dolls with celluloid heads or made completely of celluloid. Illustration **14**.

CERAMIC DOLLS: Includes *Bisque, China* and *Pottery Dolls.*

CHELSEA DOLLS: See *China Dolls.*

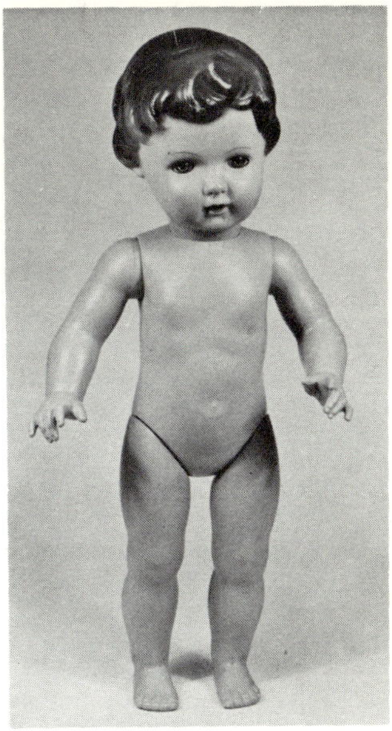

14. Celluloid doll marked on head and body with the turtle mark of Rheinische Gummi and Celluloid Fabrik Co.

CHINA DOLLS: Glazed *Porcelain Dolls.* China has a shiny surface and is translucent (not transparent) by definition. In solid china heads or when the china is very thick, it is almost impossible to see light through it. The term applies to china head dolls as well as all-china dolls. Illustrations **15A** and **15B**.

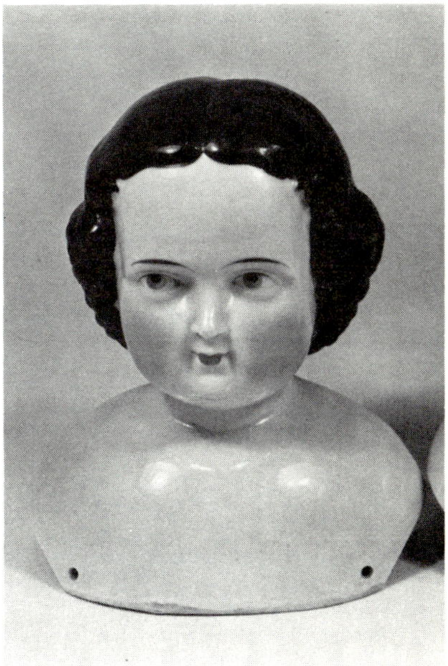

15A. China shoulder heads of the types made in the 1860s through 1880s.

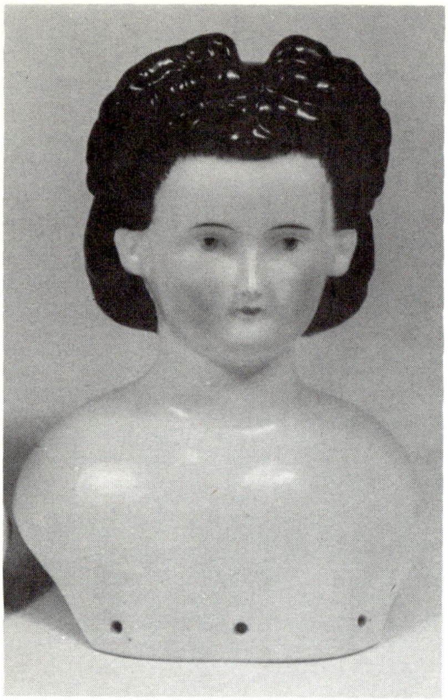

15B. China shoulder heads of the types made in the 1870s through 1880s.

CLOTH DOLLS: Dolls made of cloth, including felted fabric and treated fabrics. The cloth may be over some other material as in the case of the *London Rag Baby* which have cloth over wax. Flocking is considered to be a cloth covering. *Yarn* Dolls are not included as cloth dolls. Collectors rejected the terms Rag Dolls and Fabric Dolls. Illustrations **15** and **16**.

16. Cloth dolls; back row; Chase with bobbed hair. Nora Wellings' old lady; front row; Miss Flaked Rice made from an advertising printed fabric.

COLD-PRESS COMPOSITION: Composition pressed into molds without heat. The composition obtained by this method was used primarily prior to World War I, especially for "*Can't Break 'Em*" type head. These dolls peel rather badly even though they do not break easily. "*Can't Break 'Em*" was the *Trade Name* used by Aetna and Horsman. Illustration **17**.

17. Cold-press composition doll, "Couquette" style head made before World War I. Mouth is open-closed.

COLLECTIBLE DOLLS: Dolls made approximately 25 to 75 years ago. Technically all dolls are collectible but collectors have designated dolls older than 75 years as *Antique* and dolls less than 25 years as *Modern*. Dolls dating from between these two groups are called "Collectible." For methods of dating dolls see *Dating a Doll*. Illustration **18**.

18. Collectible dolls with bisque heads. Left, Gebrüder Heubach "Whistling Jim," about 1914. Right, Armand Marseille, marked "Just Me Registered Germany A 7/10 M," around 1930.

COMPOSITION DOLLS: A general term including *Papier-mâché* and various mixtures of materials, usually with a paper or wood-pulp base. Both *Cold Press* and *Hot Press* methods were used to produce composition. Cloth, leather, plaster, glue, whiting, vegetable matter and miscellaneous other ingredients can be used in making composition. The term applies to composition head dolls as well as all-composition dolls. *Celluloid*, *Ceramics* and *Plastics* (both hard and soft) are excluded. Illustrations **19A**, **19B**, and **19C**.

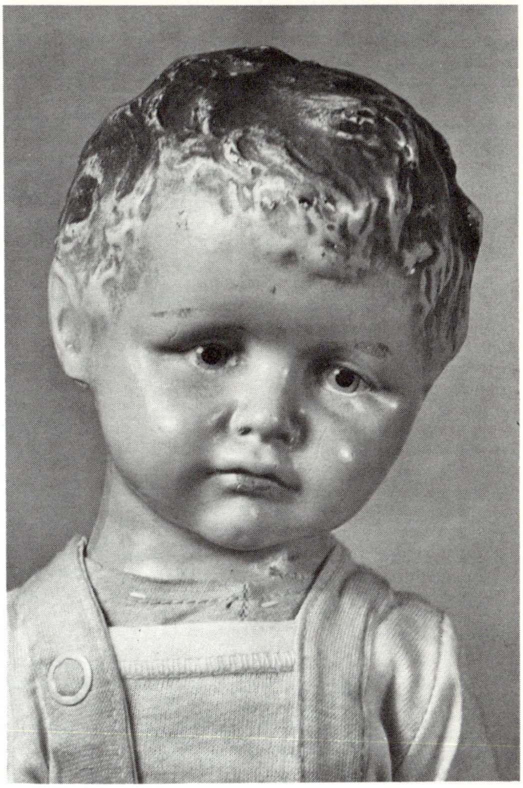

19B. Composition head of a boy doll made in early 20th century, probably by the cold-press method.

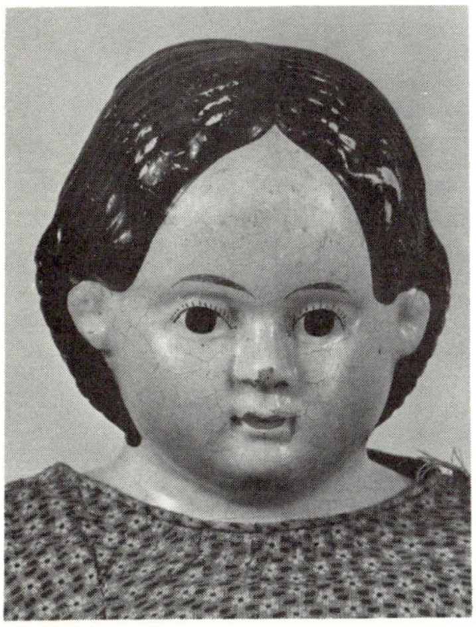

19A. Composition Greiner head with cloth reinforcement.

19C. Composition Madame Alexander Dionne Quintuplets made by the hot-press method.

CONTEMPORARY CLOTHES: Clothes not belonging to the doll originally but made either commercially or at home at about the same time as the manufacture of the doll. These clothes will fit the doll properly if the size, period and type of doll are correct. Usually it is impossible to detect whether clothes are *Original* or contemporary without proper documentation. Illustration 20.

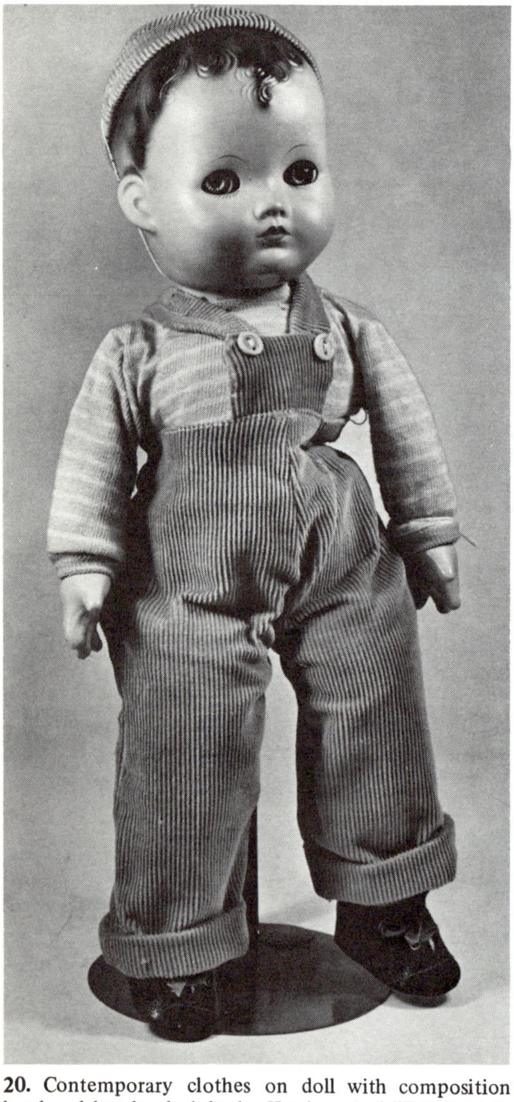

20. Contemporary clothes on doll with composition head and hands, cloth body. Head marked "F & B Made in U S A."

COOPERATIVE DOLL COMPANY: See *Joel Ellis*.

CRECHE FIGURES: These were not selected by collectors to be classified as dolls. Technically crèche figures were displayed in a Christmas nativity scene or originally were used for this purpose. Since the original purpose cannot always be determined and since the crèche figure types were sometimes utilized in the observance of religious celebrations other than the Nativity, the general type of figure usually identified with the crèche is included as a crèche figure. The French term, Crèche, is preferred by collectors to the English term, Crib; the German term, Krippe; or the Italian term, Presepio. Illustration 21.

21. Crèche figure, probably made in Naples, Italy, original clothes.

DATING A DOLL: Whatever method is used in arriving at the date should be explained so that the validity of the date can be judged. To achieve as accurate as possible a date for a doll, the following guides are listed in the order of validity as chosen by the collectors participating in the questionnaire survey. Frequently more than one guide can be applied.

1. Date from an illustration of a similar doll in a maker's or distributor's catalog, trade journal, or advertisement contemporary with the doll. The *Original Clothes* on the doll matched with such an illustration provide the most accurate means of dating a doll.
2. Date from beginning of production of pertinent type of doll. This date should be given with a plus sign (for example 1920+). It should be noted that many types of dolls were made over a long period, sometimes 50 years or more.
3. Date from *Original Clothes* where identifiable. Considerable study is often required to date *Original Clothes*.
4. Date on doll, such as patent date, the date of the founding of the firm, and so forth. The dates on dolls often refer to something many years prior to the actual manufacture of the doll. This is not a precise method of dating a doll.
5. Date from the period indicated by the artistic style. This only indicates a rough approximation of the date.
6. Dating a doll from family tradition (that is, from information by anyone except the original owner) is usually very inaccurate and is often at least a whole generation too early.

22A. Decorative figures in human form, clothes covered with shells.

DECORATIVE FIGURES IN HUMAN FORM: These were not chosen to be classified as dolls. They include figurines, wedding cake figures, dolls made into figurines by decorating them with shells, seeds and so forth. Illustrations **22A** & **22B**.

22B. Decorative figures in human form, heads and hands of beeswax, cloth bodies.

DESCRIPTION OF A DOLL: The following names are given in the order of preference by the collectors:

1. *Trade Name* used over the longest period. The *Trade Name* used at first is often short-lived (for example Patsy was first called "Mimi"). Illustration **23A.**
2. *Name* (or Names) *of Maker.* This includes maker of a part of the doll if the maker of the entire doll is not known.
3. Name of *Designer* of doll when known. Illustration
4. Material of head. Illustration **23B.**

23A. Description of a doll in order of preference:
1. Trade name "Patsy."
2. Maker, Fleischaker & Baum (Effanbee).
3. Designer, Bernard Lipfert.
4. Material, all-composition.

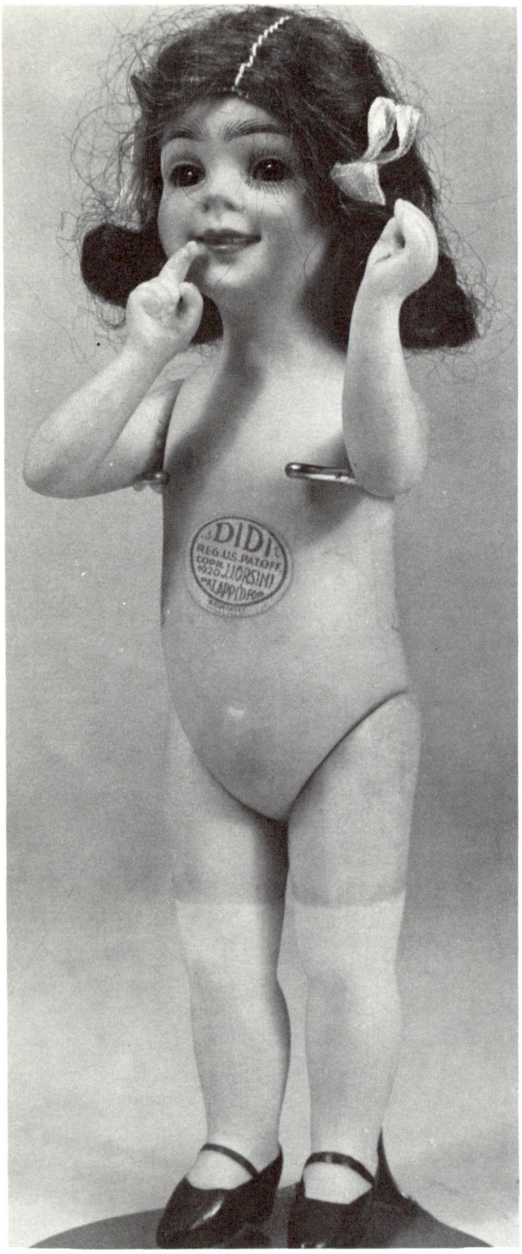

23B. Description of a doll in order of preference:

1. Trade name of "Didi."
2. Maker unknown.
3. Designer, Jeanne I. Orsini.
4. Material, all-bisque.

DESIGNER: The person who originates the design of the doll. The design can be two or three dimensional. Often the design for the doll was drawn by the designer and from this the sculptor created the model for the three dimensional doll. Illustrations **24A** and **24B**.

24. Designer: Cloth dolls made from printed fabric with Brownie designs of Palmer Cox shown in two photographs above.

DOLL: A child's plaything in human form, including unjointed figures such as *Frozen Charlottes*. All types of *Mechani-*

cal Dolls, *Peddlers*, vendors, *Fortune-telling Dolls* and *Paper Dolls* are included. The child may have been restricted in its play with certain types of dolls but they should be included nevertheless. Not included by collectors as actual dolls are figurines, action Kewpies, half figures, *Pincushion Figures*, pen-wipers, candy boxes, tea cosies, whisk brooms, wedding cake figures and other decorative doll types such as those dressed in shells or seeds. Teddy bears and other toys with animal faces are not considered to be dolls. A two (or more) faced toy with one face human and one animal face is considered to be a doll if the body is in human form and it is not a doll if the body is in animal form. Religious and ceremonial figures such as Kachinas, *Crèche Figures* and so forth are excluded. Also excluded are lobster, wishbone, pipe and similar types. Artist's and store mannequins are not included and neither are puppets or marionettes. *Lead Soldiers* and the historical figures of lead often found with them, are excluded as well as the modern plastic versions of these toys. Illustration **25**.

25. Dolls originally childrens' playthings: Left, China Frozen Charlotte. Right, peddler doll made of leather and cloth with simulated bolts of fabric on the head.

DOLLS' HOUSE DOLLS: Any doll that can be used in a dolls' house regardless of size or material. Illustration **26**.

26. Dolls' house dolls with bisque heads and limbs, child doll is all-bisque. Left to right, a lady, a chauffeur, a little boy, a gentleman.

DREAM BABY: New-born baby type doll with bisque head made by Armand Marseille, cloth or composition baby body, composition, celluloid or rubber hands. The *Trade Name* was "My Dream Baby." Illustration **27**.

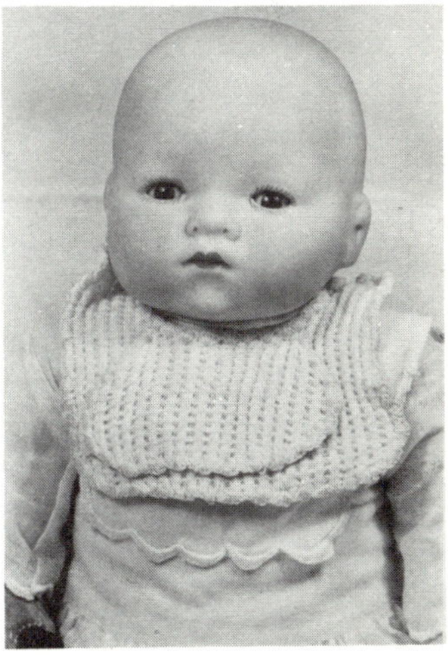

27. Dream Baby, bisque head, cloth body, celluloid hands, closed mouth (Dream Babies can also have open mouths). Mark on head, "A. M. Germany 341. 13."

DRESDEN DOLLS: See *Parian-Type Dolls*.

DUTCH DOLLS: See *Peg-Wooden Dolls*. The term Dutch Dolls was rejected by present-day collectors because it is probably a contraction of the word "Deutsche" meaning German. The term "Dutch Doll" might mislead people into thinking that these dolls had a Holland-Dutch origin instead of a Germanic origin. Actually most of them appear to have been made in the Austrian Tyrol. Illustration 28.

28. Dutch dolls, rejected term. The wooden dolls pictured are large versions of peg-wooden dolls and have ball and peg joints.

EARLY MOLDED HAIR PAPIER-MACHE: A type of doll sometimes called "*Milliners' Model*" (a term not chosen by the majority of collectors). These dolls with papier-mâché molded heads, unjointed kid torsos and wooden limbs were produced primarily in Germany from 1810-60. Occasionally these are found with cloth bodies including cloth limbs, but sometimes having wooden limbs. Prior to the introduction of the term "*Milliners' Model*" (see also) in the early 1950s, this type of doll was called "Antique Papier-mache" by collectors. Since the word Antique covers a longer period of time now than it did then, the word Early was substituted to differentiate these dolls from the later papier-mâché dolls that collectors now call *Antique*. The collectors who participated in the survey were practically equally divided in their choices on the questionnaire for *Milliners' Model*, Coiffured Head and Varnished Head. Thus two-thirds of the collectors in the survey were opposed to any one of these three terms. Since the distinctive characteristic of these dolls is their molded hair, they are best described as Early Molded-Hair Papier-Mâché. Illustration 29.

29. Early molded hair papier-mache dolls with kid bodies and wooden limbs.

ETHNIC DOLLS: Dolls representing people of various races or nationalities. This would include *Oriental*, *Indian*, Negro, Polynesian, Eskimo and so forth. These dolls have the physical characteristics of the people represented and may or may not have been made by the people depicted. Sometimes the physical characteristics are not apparent when the dolls are made by natives. Also see *Regional Costume*. Illustrations **30A** and **30B**.

30A. Ethnic doll representing an African, bisque head, composition body. Marked: "Heubach Köppelsdorf 459. 13/0 Germany."

30B. Ethnic dolls, bisque heads. Left, German interpretation of an Oriental, doll marked "Simon & Halbig Germany 4." Right, Negro doll marked "Germany Simon & Halbig S & H 6."

FASHION-TYPE DOLLS: Dolls originally dressed or intended to be dressed in fashionable or *Regional Costume*. These dolls are dressed as adults or older children. The majority have bisque heads and shapely kid bodies. Illustration **31**.

31. Fashion-type dolls, bisque heads, kid bodies. Left, dressed in original regional costume. Right, dressed as a young girl, original clothes.

FELT DOLLS: Dolls with felt heads or made completely of felt (for example Lenci dolls). The felt may or may not be treated. Illustration **32**.

32. Felt doll made by Lenci, original regional clothes.

FLANGE NECK: The bottom of the head piece and the top of the shoulder piece fit together with a flat surface. Sometimes there is a flange around the edge of the joint. This type of joint is sometimes referred to as a "cup and saucer" joint but this term was not chosen by collectors. Illustration 33.

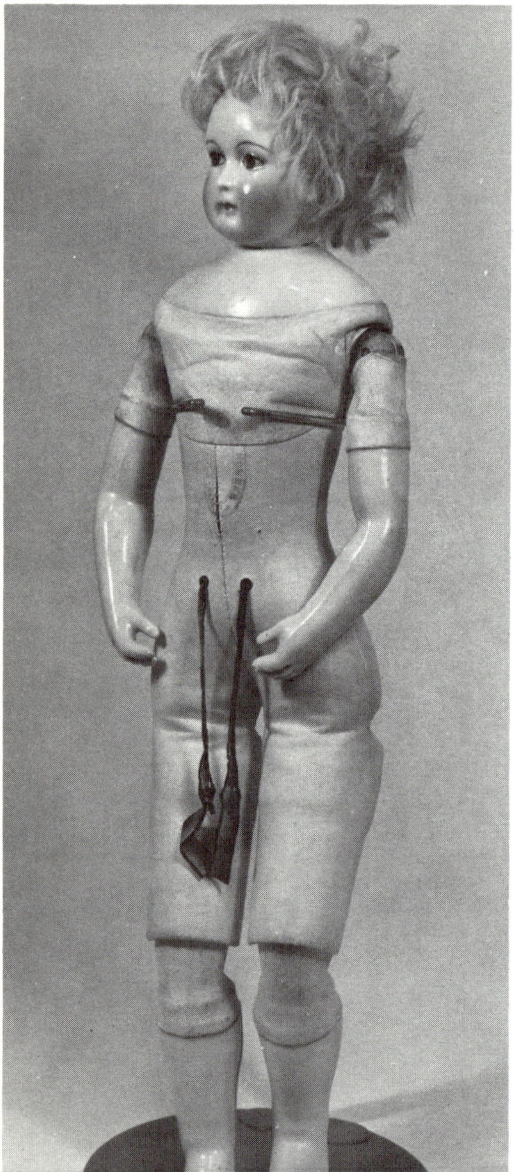

33. Flange neck on china head, marked Rohmer kid body with china limbs.

FIGURINES: See *Decorative Figures.*

FOLK DOLLS: See *Primitive Dolls.*

FOREIGN DOLLS: See *Regional Dolls.*

FORTUNE - TELLING DOLLS: Dolls with fortunes written on the segments of their skirts or holding a pointer to indicate the fortune. See also *Dolls.* Illustration **34.**

34. Fortune telling doll, bisque head, skirt made of paper slips with fortunes on them.

FRENCH BISQUE: A misleading term sometimes used to designate a good grade of bisque. It should not be used in this sense by collectors, if it does **not** relate to the country of origin of the doll. See *French Dolls.*

FRENCH DOLLS: The majority of the collectors who answered the questionnarie wished the term "French Doll" to be applied only to dolls made entirely in France. Under this definition there will be a severe limiting of the number of dolls that can be called "French." According to this definition any doll with an unmarked or unrecognizable marked head cannot be called "French" or given any other definite country of origin. (Two bisque heads with identical contours, one with a French mark and one with a German mark have been reported. Thus the country of origin must be verified by the mark.) See also *French-Type Dolls.* Illustration **35**.

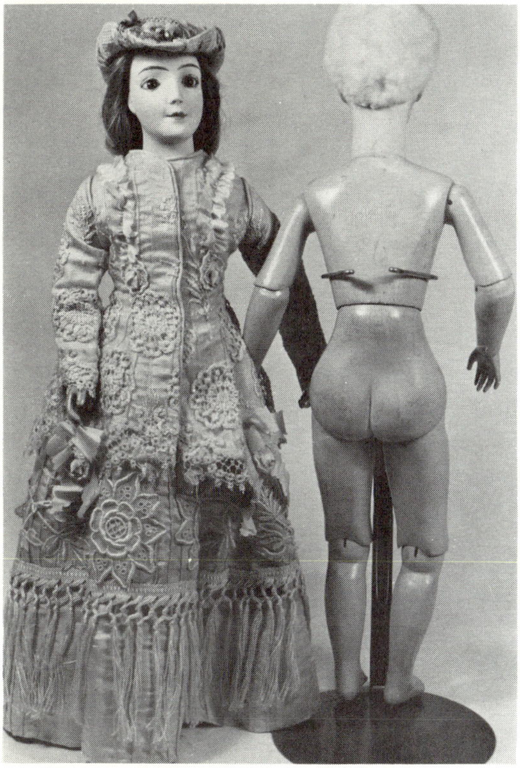

35. French doll, bisque head marked, "Huret," body marked "Huret 68 Rue de la Boetie."

FRENCH FASHION: See *Fashion-Type Dolls* and *French Dolls.*

FRENCH-TYPE DOLLS: Dolls made partially in France including dolls that cannot be proven to have been made entirely in France are termed "French-Type Dolls." Illustrations **36A** and **36B**.

36A. French-type doll, bisque head marked "S & H 10", composition body labeled "Bébé Jumeau. Diplome d'Honneur." This doll could also be called a German-type doll.

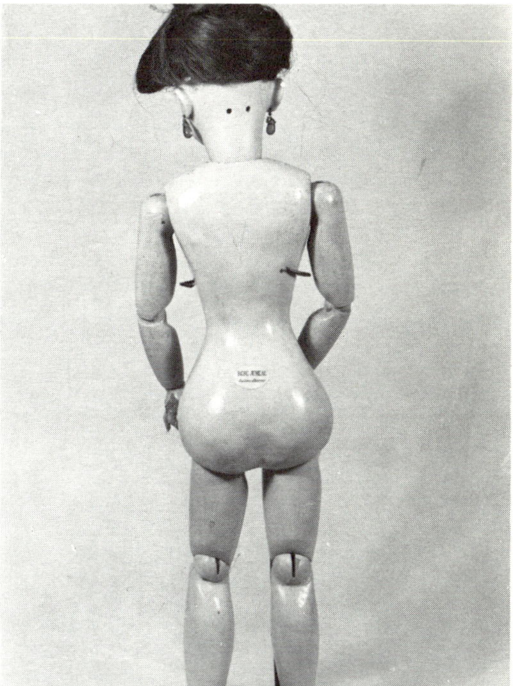

36B. French-type doll, back view of 36A.

FRENCH-TYPE ALL-WOODEN: The *Trade Names* for these dolls are "Bebe Tout En Bois" and "Bebe Tout Bois." They were sold in Paris stores from about 1900-14. One of these dolls has been found with its original label on the body which states "Made in Germany." Illustration **37**.

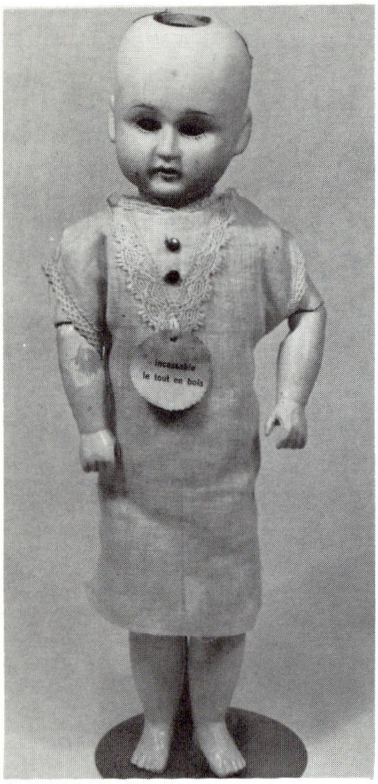

37. French-type all-wooden doll possibly made in Germany, distributed by French stores.

FRENCH-TYPE COMPOSITION: Composition head dolls made primarily in France. These dolls were made from about 1790 to 1880. The early ones have part or all of their torsos of composition or *Molded Cardboard*, later the bodies were made of pink kid and the later ones of white kid or cloth. The dolls have bald heads, often with black pates. Usually they have wigs or a headdress. Illustration **38**. See below.

FROZEN CHARLIE: Rejected by collectors. See *Frozen Charlotte*.

FROZEN CHARLOTTE: An unjointed doll generally made of bisque or china, occasionally made of other materials. A *Trade Name* was "Bathing Baby." Illustration **39**.

39. Frozen Charlottes, all-china dolls with and without molded clothing.

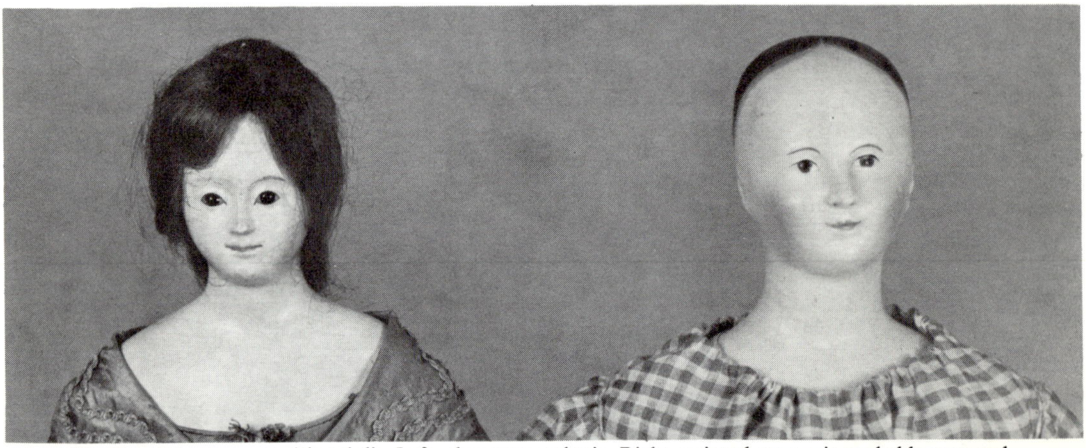

38. French-type composition dolls. Left, glass eyes and wig. Right, painted eyes, wig probably removed.

FUR-CLAD DOLLS: Dolls dressed completely in fur. The fur is usually white and attached to the body and head. These dolls were made especially around 1900 when there was widespread interest in the arctic region. The names Eskimo Dolls, Baby Bunting, Rupprecht and so forth were not chosen by collectors. Illustration **40**.

40. Fur-clad dolls dressed in original clothes, carved wooden heads.

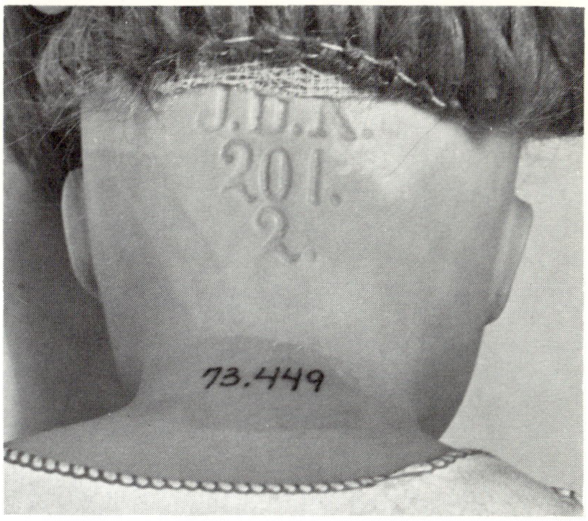

41B. Close-up of Kestner mark on bisque head.

GERMAN DOLLS: Dolls made entirely in Germany. Illustrations **41A** and **41B**.

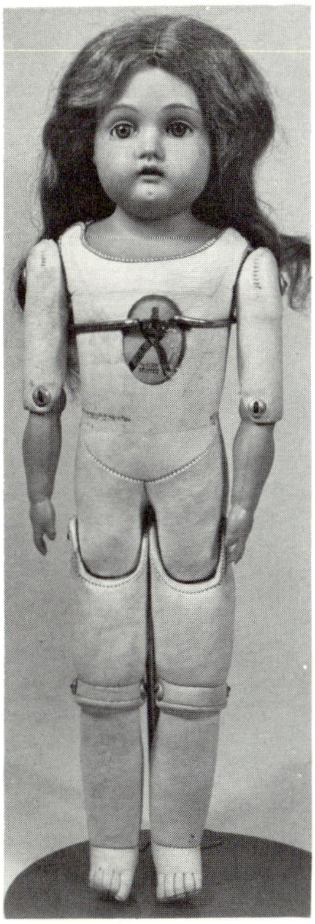

41A. German doll, bisque head marked "J D K 201 3", kid body with Kestner label having the crown and streamers trademark.

GERMAN-TYPE DOLLS: Dolls made partially in Germany, including dolls that cannot be proven to have been made entirely in Germany. Illustration **42**.

42. German-type doll, Armand Marseille bisque head, composition body with original box labeled "L. C. 'Hush-A-Bye' Doll Assembled and Dressed in Australia."

GOOGLY EYES: Eyes glancing to the side. The eyes can be painted or glass inset. The *Trade Name* was "Roguish eyes." Illustration **43**.

43. Googly eyes, bisque head marked "Germany A 3/0 M," glass eyes, composition body, original clothes.

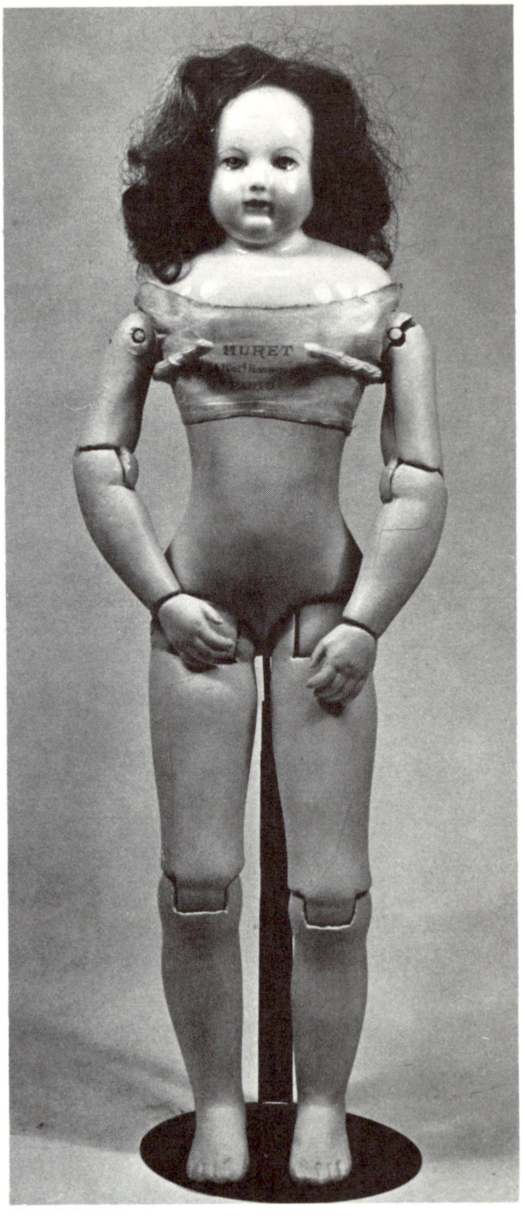

44. Gutta Percha, this body appears to be made of gutta percha, china head, Huret mark on chest.

GUTTA PERCHA: Gutta Percha dolls are not made from pure gutta percha which is pinkish white in color, very brittle when hard and will not float. It is currently used for golf balls and insulators. In the nineteenth century various synthetic compounds were called "Gutta Percha." When gutta percha itself was used, resins and other substances were added. Gutta percha should not be confused with rubber which is dark in color and lighter in weight. Illustration **44**.

HALF DOLLS: Dolls with molded upper torsos and cloth lower torsos. The upper parts of the limbs are also cloth. These dolls are not to be confused with half figures used for *Pincushions* and similar objects. Illustration **45**.

45. Half dolls, bisque upper torsos and limbs, cloth lower torsos.

HALF FIGURE: See *Pincushion Figure*.

HATTED DOLLS: Dolls with molded hats, bonnets or headdresses including *Snow Babies*. Illustration **46**.

HARD PLASTIC: A special kind of composition developed for dolls after World War II. It is a hard material that gives a sharp delineation to the features.

HOT-PRESS COMPOSITION: Composition pressed into molds with heat. This method was used for dolls during and after World War I. One of the principal ingredients was wood-pulp. The composition is light brown in color on the inside of the heads. Illustration **47**.

47. Hot-press composition head and hands, label reads: "Lullabye Baby, Genuine Madame Hendren Doll," original clothes.

46. Hatted dolls, left, molded baby bonnet, head marked "R A." Right molded pill-box hat.

IDENTIFICATION OF A DOLL: A clearly marked doll can be identified provided the mark is known. Dolls in *Original Clothes* that clearly resemble illustrations in catalogs or advertisements contemporary with the doll might be identified. It is necessary that the clothes be original and that the costumes are compared in detail.

An unmarked doll or a doll marked only with numerals and/or a single letter usually cannot be identified, except for the Kestner size series with a letter and numeral. Dolls in marked boxes cannot be identified unless it is absolutely certain that the box is original to that particular doll. Dolls that closely resemble marked dolls or illustrations of marked dolls cannot necessarily be identified as the same doll. Similar appearing dolls are usually of the same period. Also see *French Doll* and *Maker*. Illustration **48**.

48. Identification of doll can be made by the mark on the bisque head which is "Germany 390 A $5/0^{X}$ M," an Armand Marseille mark. The doll is originally dressed as a milk maid of Flanders as indicated by the label, "Laitière Flamande."

INDIAN DOLLS: Dolls made by members of the Indian races inhabiting the Americas. Also see *Ethnic Dolls* and *Regional Costume*. Illustration **49**.

49. Indian dolls in original clothes. Left, Navajo made of cloth. Center, wooden Kachina. (This is not a doll because the head is not in human form.) Right, Cheyenne made of cloth and leather.

INSERTED HAIR: Hair or hair substitute inserted into the scalp of a doll. The doll may be made of any type of material. See also *Montanari*. Illustration **50**.

50. Inserted hair on vinyl doll known as "Mrs. Beasley."

INTAGLIO EYES: Painted eyes having semi-spherical impressed or incised pupils. Illustrations **51A** and **51B**

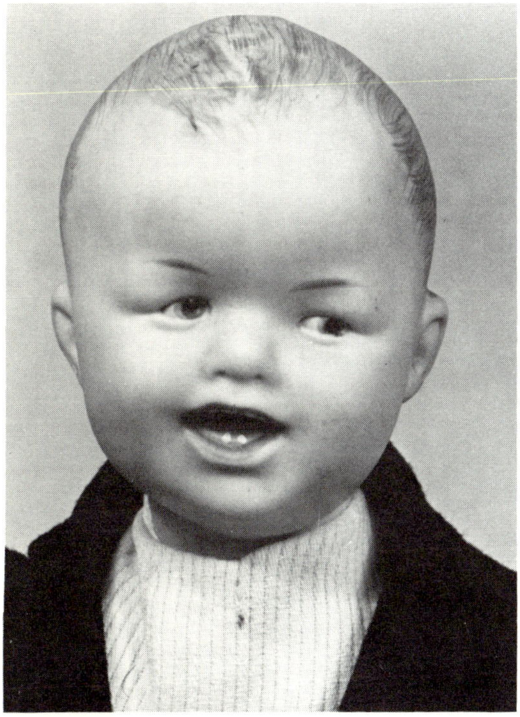

51A. Intaglio eyes on bisque head made by Gebrüder Heubach in Germany, mold number 31 01.

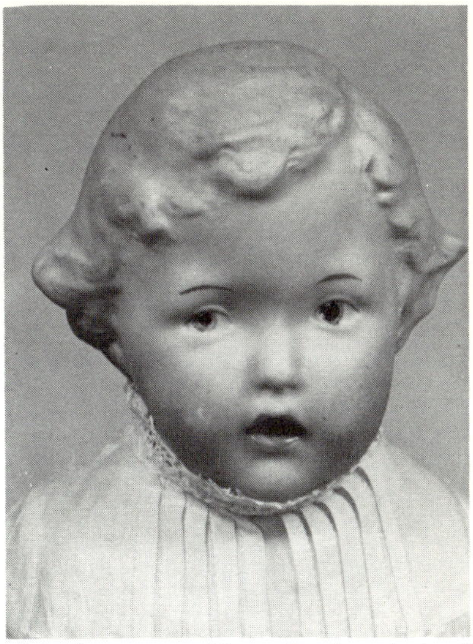

51B. Intaglio eyes on bisque head made by Gebrüder Heubach, mold number 31 39.

IVORY DOLLS: See *Bone Dolls*.

JOEL ELLIS DOLLS: Dolls made by the Cooperative Doll Company in Springfield, Vermont around 1873. These were all-wooden dolls except for metal hands and feet. Illustration **52**.

52. Joel Ellis doll, all-wooden except for metal hands and feet.

JOINTED DOLL COMPANY: Made wooden dolls with composition heads and usually metal hands and feet in Springfield, Vermont. Many of these dolls have a black band around their waist with dates for the patents of Martin, Sanders and Johnson. When the black band has been lost it is difficult to differentiate these dolls from the *Mason and Taylor Dolls.* Illustration **53.**

53. Jointed Doll Company's doll with composition head, wooden body, metal hands and feet. Black band around waist has identifying patent information.

KEWPIES (Registered Trademark): Copyrighted dolls made from the Kewpie designs of Rose O'Neill. Some part of a Kewpie must be jointed in order to qualify as a doll, otherwise it is an "action Kewpie" and not classified as a doll. This was based on the decision of the collectors who participated in the questionnaire survey. Illustration **54.**

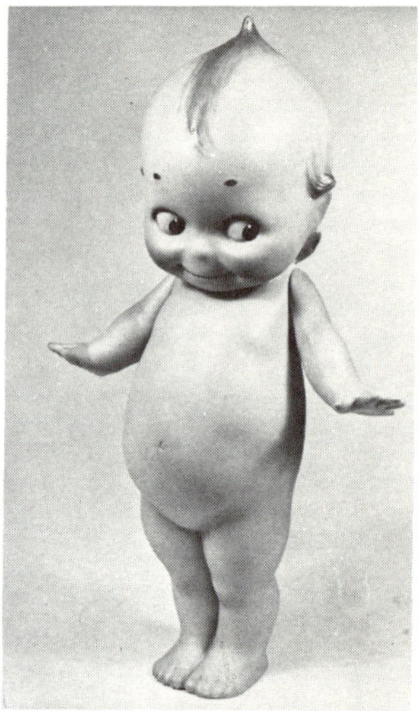

54. Kewpie copyrighted by Rose O'Neill, all-bisque version © Joseph L. Kallus.

KID BODY: All or the majority of a doll's body made of kid or leather. Also see *Leather Dolls.* Illustration **55.**

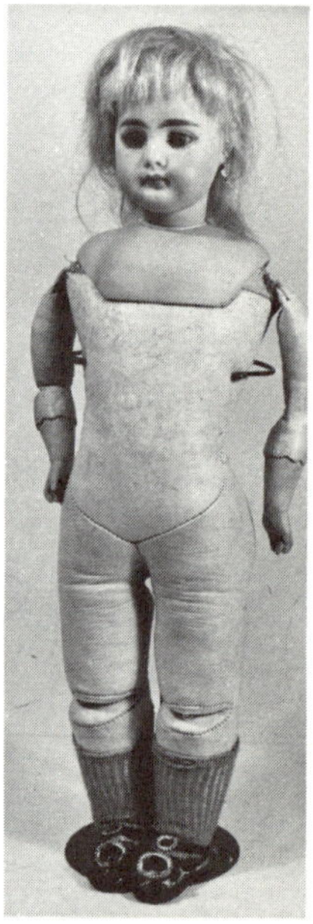

55. Kid body, bisque shoulder head with swivel neck, bisque hands, no mark.

LADY DOLLS: Dolls with bodies having the contours of an adult female. Illustration **56**.

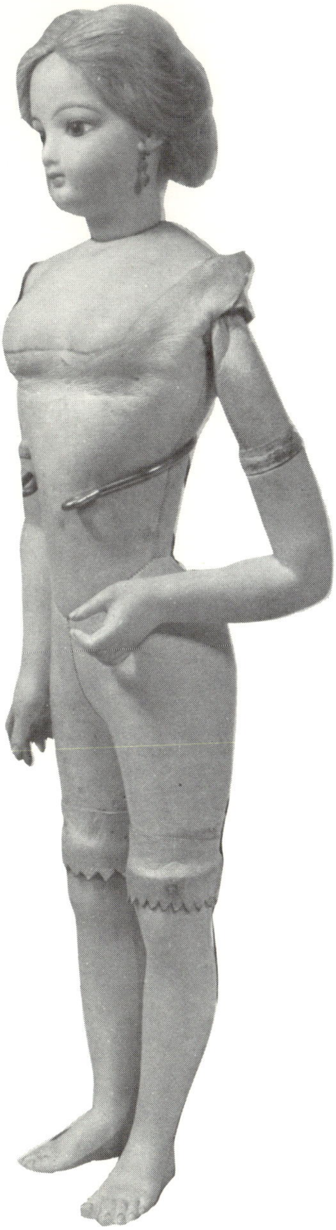

56. Lady doll with bisque head and limbs, kid body in the shape of an adult female.

LAYETTE: A *Wardrobe* (see also) for a baby doll. Illustration **57**.

57. Layette for a composition Effanbee "Patsy Baby"; marked on head. Doll and layette in original basket.

LEAD SOLDIERS AND FIGURES: Not classified as dolls by collectors. This includes the historical type lead figures often found with the lead soldiers. Illustration **58**.

58. Lead figures representing Henry VIII and four Queens, made in England. These are not classified as dolls.

LEATHER DOLLS: Dolls with leather heads or made completely of leather. The leather may be whole or split. It may have a smooth finish or a suede finish. The skin may be kid, goat, calf, lamb and so forth. Illustration **59**.

59. Leather doll patented in 1903 by Gussie Decker. It laced up the front.

LONDON RAG BABY: A doll with a mask face made of wax covered with thin cotton cloth. This is classified as a *Cloth Doll.* Illustration **60**.

60. London Rag Baby with wax-mask face covered with thin muslin.

LONG-FACED JUMEAU: Name given to a doll with a special type of long bisque face having a closed mouth. The ball-jointed composition body but not the head, is marked "Jumeau." The term "Cody Jumeau" for this type of doll was not chosen by collectors. Illustration **61**.

61. Long-face Jumeau, unmarked bisque head, marked Jumeau composition body.

MAKER: The person or company who made the head of the doll or if that is not known, the person or company who made the body of the doll. If neither of these is known, the assembler is called the maker even though the various parts were made by different persons and/or companies. Most dolls had a variety of companies that produced the different parts. For example Borgfeldt hired the designer, contracted with the makers of the various parts and assembled the dolls. In the case of the Bye-Lo heads several different porcelain factories made these heads. A number alone on a doll practically never indicates the maker since most numbers were used by more than one maker or assembler. However, the arrangement of the numbers and/or letters sometimes is a clue to the name of the maker or assembler. Also see *Identification of a Doll*. Illustration **62**.

62. Maker of this doll is described as Gebrüder Heubach although only the bisque head is made by Heubach; sunburst Heubach mark with "Germany 1." Original regional costume.

MAKER'S NAME: The name of the maker of the head takes precedence over the name of the maker of the body or the company that assembled the doll.

MAMA DOLL: Doll with a plump body having a squeak box in it. The swinging legs are constructed so that when the doll is held and led forward it simulates walking but there is no walking mechanism. Illustration **63**.

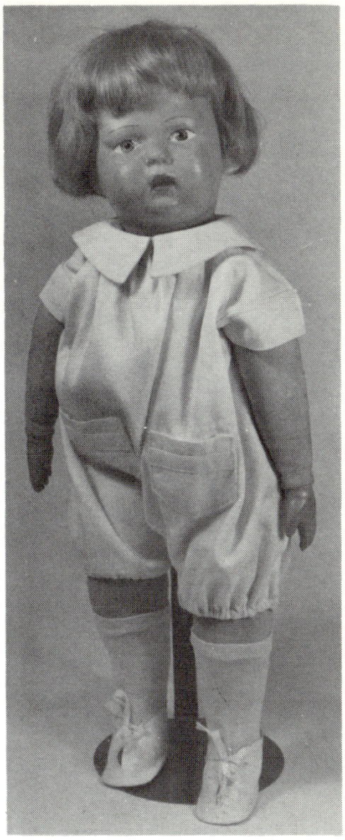

63. Mama doll, wooden head and hands, cloth body, made by Schoenhut.

MANIKINS: See *Mannequins*.

MANNEQUINS: Artist and store mannequins are not classified as dolls in accordance with the collectors' preference. Illustrations **64A** and **64B**.

64A. Mannequin for an artist's use, all wood, completely jointed. Mannequins are not classified as dolls.

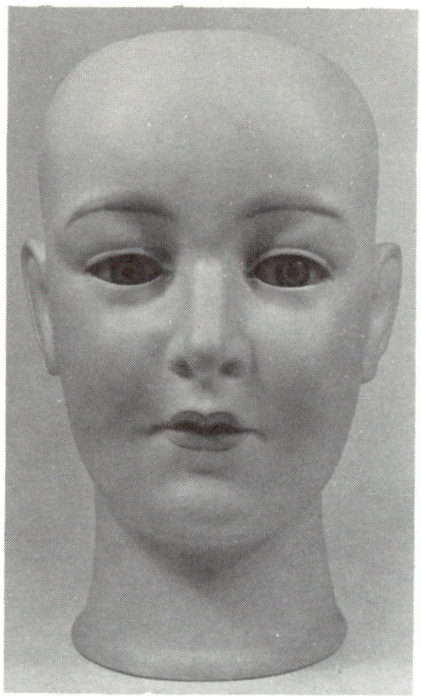

64B. Mannequin's life-size bisque head, probably used for display purposes.

MARKS: It takes a real expert to identify some porcelain marks. For example there were many manufacturers who used a crossed swords mark similar to *Meissen.* Wallendorf appears to have used not only the crossed swords, but also the Berlin W and the Nymphenburg crosshatched mark. The Rudolstadt mark of a crown over a N is almost identical to the Capo di Monte mark. It is known that Rudolstadt made dolls. Some bisque head marks are often mistakenly read. Many times "Made in" is read as "Madeline."
Illustrations **65A**, **65B**, **65C**, **65D**, **65E** and **65F**.

65A. Marks used by Wallendorf in Thuringia that are almost identical to marks used by Nymphenburg, Berlin and Meissen.

65B. Mark of Capo di Monte as imitated by Rudolstadt of Thuringia.

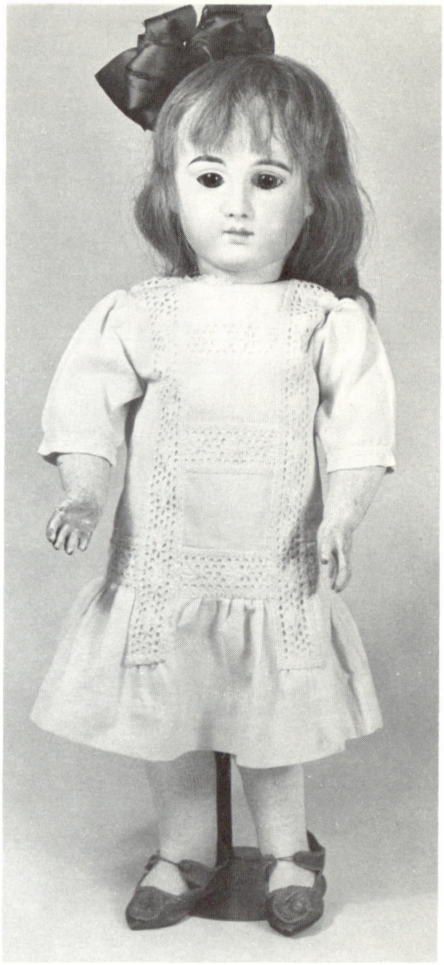

65C. Marks on this bisque head, composition-bodied Schmitt doll are shown in **65B. C.** and **D.**

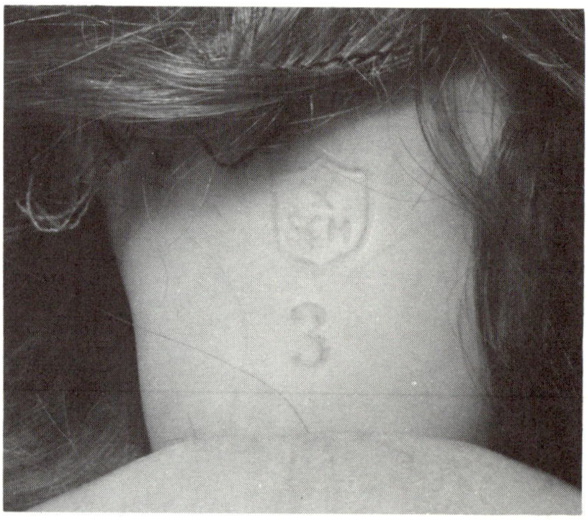

65D. Mark incised on bisque head of Schmitt doll.

65E. Mark as a label on composition Schmitt body indicating that it was distributed by Guyot.

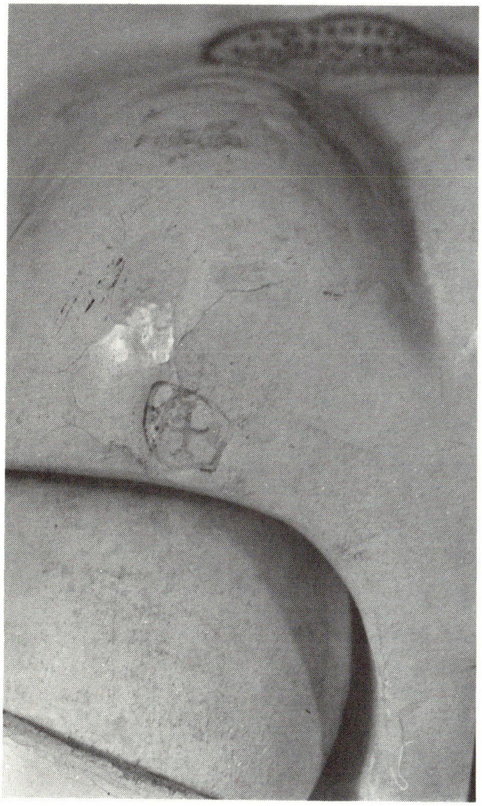

65F. Mark on buttock of Schmitt doll. This mark stamped on the composition body is the same as the mark incised on the bisque head.

MAROTTE: Doll head and upper torso on a stick that can be twirled or agitated and usually music accompanies the movement. The end of the stick often has a whistle. This type of doll is also known as a "Folie" or "Folly." Dolls' heads on sticks that cannot be twirled or agitated are not included as marottes. Illustration **66**.

66. Marotte plays a tune when twirled, bisque head unmarked.

MASON AND TAYLOR DOLLS: Wooden dolls with composition heads and metal or wooden hands and feet made by D.M. Smith & Company in Springfield, Vermont. One type of doll was called "Witch and Wizard." Also see *Springfield Wooden Dolls*.

MATERIAL OF DOLL: The material of the head and not the body, is used to designate the type of doll, such as bisque, china, wooden and so forth. Illustration **67**.

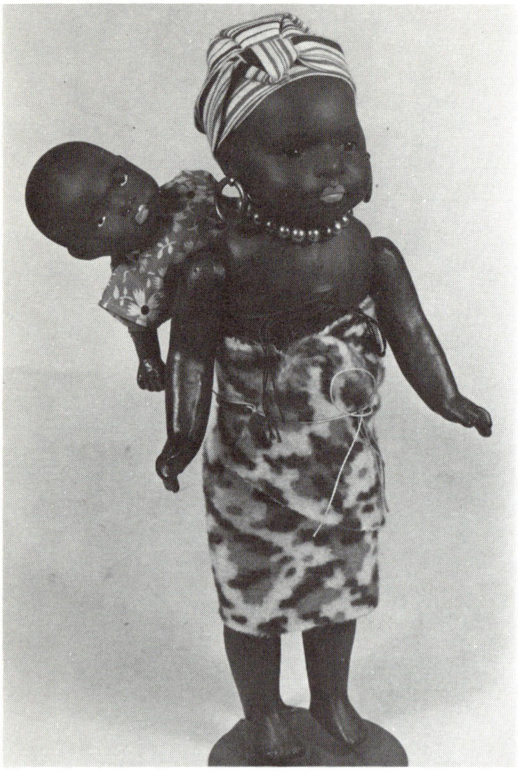

67. Material of dolls is designated as bisque because the Heubach of Köpplesdorf heads are bisque. The bodies of these ethnic dolls are composition.

MECHANICAL DOLLS: Dolls performing simple or complicated functions by means of mechanisms. They can be attached to a base or stick such as a *Marotte* (see also). Sound may or may not be included. The mechanism may be as simple as that used to turn the doll's head when it walks or to say "Mama" and "Papa" when strings are pulled. *Mama Dolls* (see also) and dolls with eyes that move by gravity are not considered to be mechanical dolls. Illustration **68**.

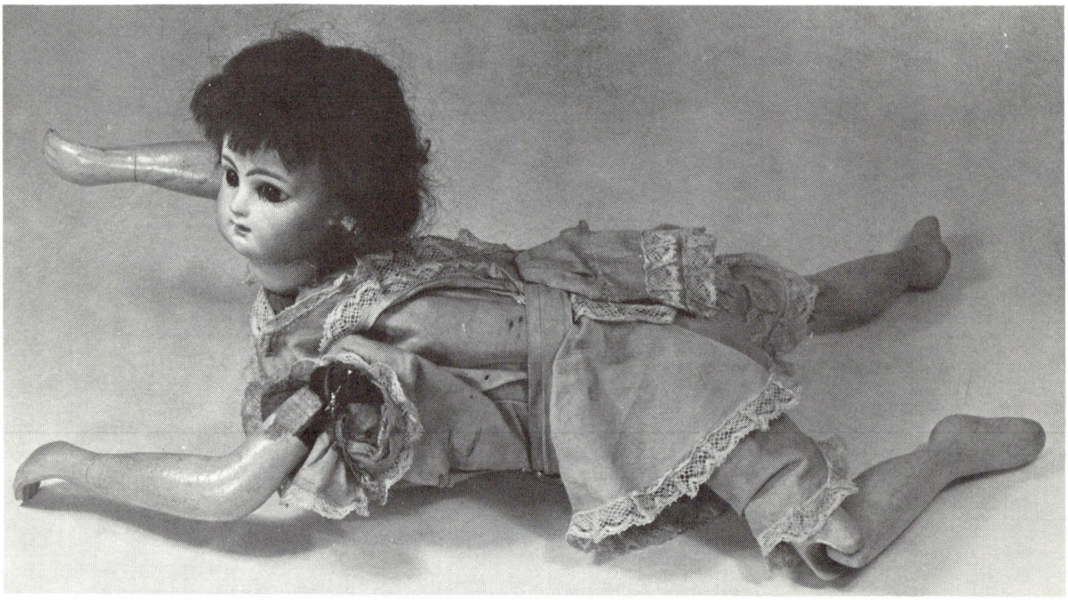

68. Mechanical doll known as the "Ondine" swimming doll, bisque head.

MEISSEN DOLLS: The authentic Meissen mark must be on the doll. It must be remembered that many porcelain manufacturers used a crossed swords mark, similar to those associated with Meissen. Also see *Marks*. Illustrations **69A** and **69B**.

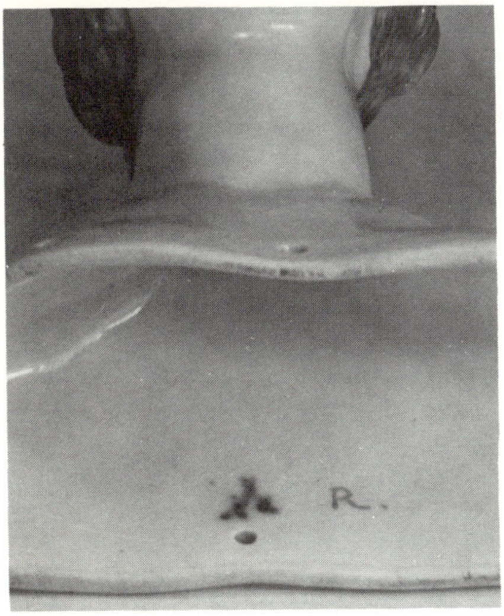

69A. Meissen crossed swords mark on doll's head shown in **69B**. Many porcelain factories used a crossed swords mark making it diffucult to determine an authentic Meissen head.

69B. Meissen china shoulder head with crossed swords mark, see **65A** and **69A** for discussion of the possibility that this might not be a genuine Meissen head.

METAL DOLLS: Dolls with metal heads or made completely of metal. The metal may be tin, brass, pewter, steel, silver or other metallic substances. Illustration **70**.

70. Metal dolls' head marked "Minerva" with its original box. Head made by A. Vischer & Co.

MILLINERS' MODEL: The majority of the collectors voted against the use of the term "Milliners' Model." These dolls were toys for children and not used by milliners. The term "Milliners' Model" was originated by Eleanor St. George in the early 1950s. See *Early Molded Hair Papier-Mâché.*

MINT: Dolls in original condition and appearance just as they were when new. This applies to the clothes and accessories as well as the doll itself. A *Restored Doll* cannot be called "Mint." Illustrations **71A** and **71B**.

71A. Mint condition of a doll with a peg-wooden body and wearing clothes of the mid-19th century.

71B. Mint condition of a Lenci doll with felt head and Tyrolean regional costume, around 1940.

MODERN DOLLS: Dolls made less than 25 years ago. For method of dating dolls see *Dating a Doll*. Illustrations **72A** & **72B**.

72A. Modern doll with vinyl head marked H. probably for Horsman.

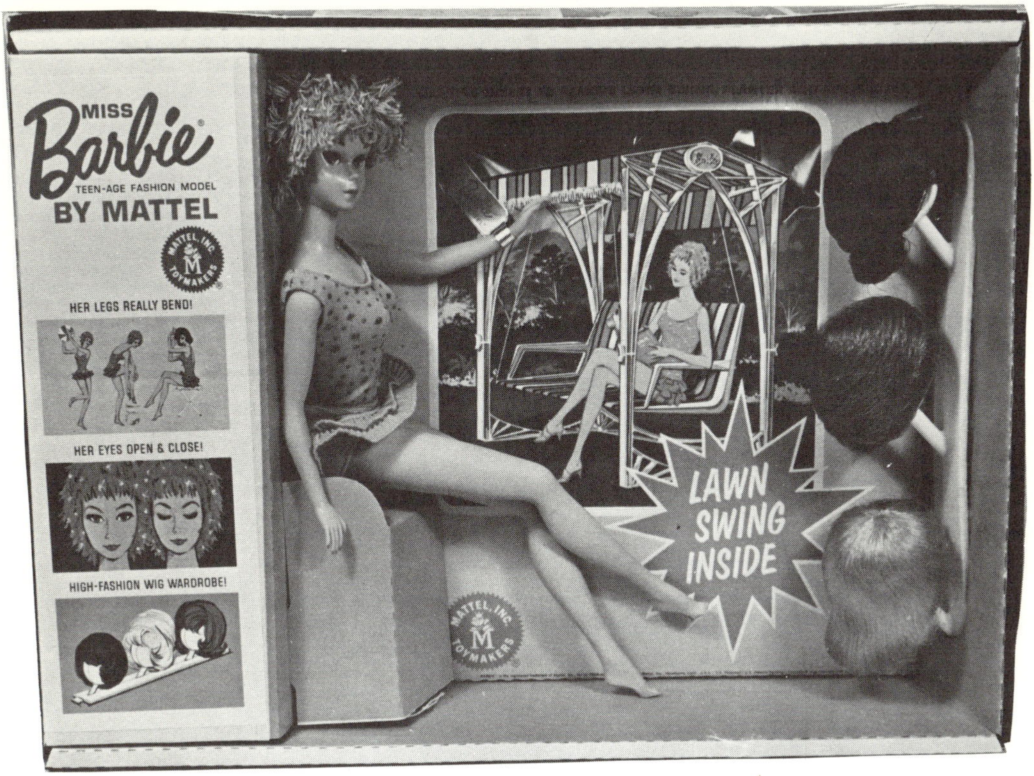

72B. Modern doll "Miss Barbie," the "Teen-age fashion model" made by Mattel, shown in its original box with a wig wardrobe.

MOLD: The form into which slip is poured or malleable material is pressed in order to produce parts of a doll. The molds in a negative or intaglio form are made of a variety of materials, among them plaster, sulphur, metal and so forth. These molds are usually derived from a master mold which is a positive form and has contours similar to that of the doll itself.

MOLDED: Contours of doll formed by pouring or pressing into or onto a mold. This also refers to molded hair and/or molded clothes. Illustration 73.

73. Molded doll's head of bisque made by pouring slip or pressing into a mold.

MOLDED CARDBOARD: Material used especially in France, for making dolls (or dolls' heads) at the end of the 18th century and the first half of the 19th century. Molded cardboard was used much later just for dolls' bodies. These were probably made primarily in Germany. Illustration **74**.

74. Molded cardboard used for dolls' heads. This pair probably made in France in the late 18th century; original clothes.

MONTANARI: Poured wax dolls with inserted hair and having a Montanari mark. Unmarked wax dolls cannot be called Montanari even when they have *Inserted Hair*. See *Montanari-Type*. Illustration **75**.

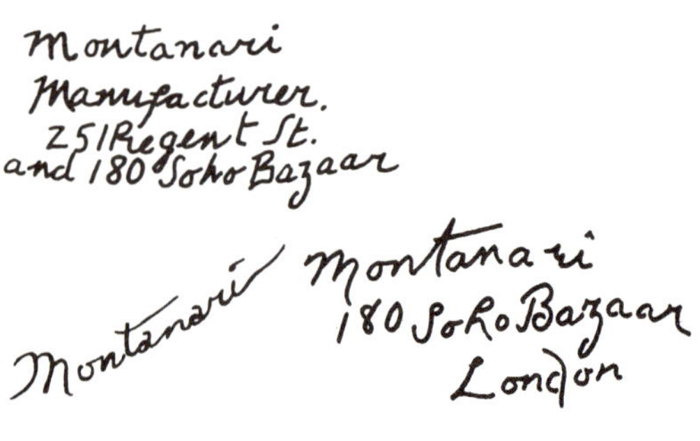

75. Montanari marks found on wax dolls. These marks are written in ink on the cloth bodies.

MONTANARI-TYPE: Unmarked poured wax dolls with *Inserted Hair* (see also *Wax With Inserted Hair*). Illustration **76**.

76. Montanari-type dolls with poured wax heads and arms, inserted hair, no marks.

MOTSCHMANN-TYPE BODY: A doll's body nearly always with cloth upper torso and a hard substance for the lower torso, hands and feet. The arms and legs are often wooden tubes that provide floating joints. Motschmann patented this type of doll in Germany in 1857 but they were also made elsewhere. A squeak box was usually placed in the upper torso and these dolls were generally dressed as babies originally. Illustration **77**.

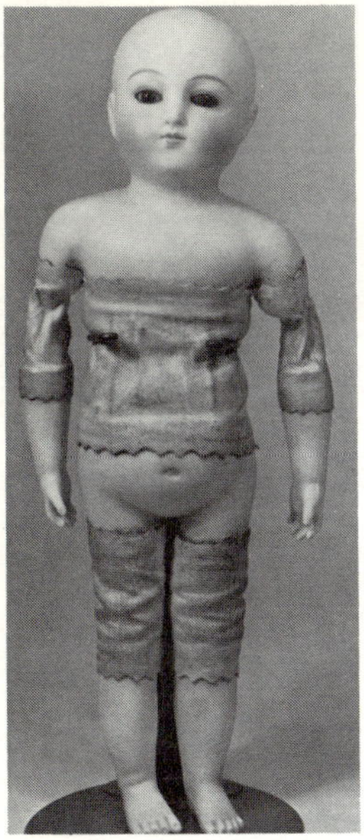

77. Motschmann-type body with bisque head and body parts. Upper torso and upper limbs of cloth.

78B. Multi-face bisque head crying, see **78A** and **78C**.

MULTI-FACE DOLLS: Dolls having two or more faces on the same head or with additional mask faces.
Illustrations **78A**, **78B**, and **78C**.

78A. Multi-face bisque head smiling, see **78B** and **78C**.

78C. Multi-face bisque head sleeping, see **78A** and **78B**.

78D. Multi-face doll with wax masks representing infancy, youth and old age, also see 78E.

MULTI-HEAD DOLLS: Dolls having two or more heads. Two heads can be attached to a *Topsy-Turvy* type body or more than two can be attached at the neck, one at a time. Illustration **79**.

79. Multi-head doll with three composition heads, composition body stamped "D R G M 16 R47." Around 1890s.

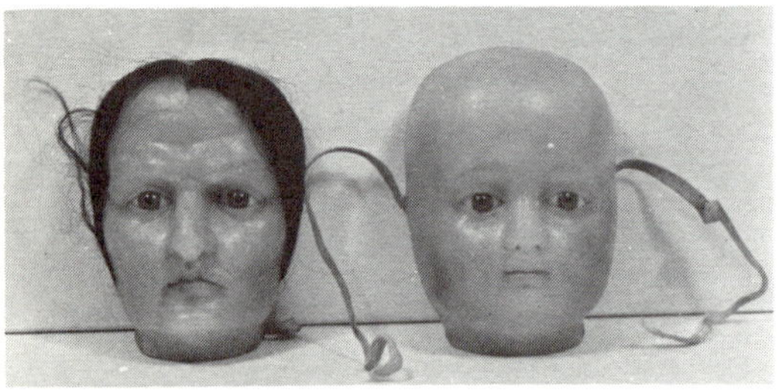

78E. Multi-face masks belonging to doll shown in **78D**. Strings that tie the mask onto the head are seen on right.

NAME DOLLS: *Trade Names* and/or names on the dolls themselves such as Marion, Patsy and so forth, should be used as the name of the doll. However, names recently applied to old dolls such as Jenny Lind, Empress Eugenie, Godey, Gypsy Jumeau, *Portrait Jumeau* and so forth should not be used in writings, exhibit classifications and other formal purposes. Frequently these names are of regional use only and not known specifically to the majority of collectors. Illustration **80**.

80. Name dolls, china shoulder heads with the names: "Marion," "Dorothy," and "Edith" reading left to right. Trade name for these dolls was "Pet name."

NAME OF MAKER: See *Maker's Name.*

NANKEEN DOLLS: *Trade Name* for dolls with bodies of cotton cloth and usually china or bisque heads and lower limbs. Collectors prefer to use the term cloth body dolls.

ONE-OF-A-KIND: Dolls can be homemade *Cloth Dolls* or early woodens such as *Queen Anne* and *Peg-Wooden Dolls.* Dolls made in a mold can practically never be referred to as "one-of-a-kind." However, since nearly all dolls are painted by hand, there is a slight variation in appearance but this does not constitute its being "one-of-a-kind. Illustration **81**.

81. One-of-a-kind homemade ethnic cloth doll.

OPEN-CLOSED MOUTH: The lips are parted on the doll but there is no real opening between the lips cut into the head. Illustration **82**.

82. Open-closed mouth on bisque head, often the lips are wider apart and sometimes teeth are molded but without any opening.

ORIENTAL DOLLS: Dolls with Oriental features. These dolls may or may not be made by the people represented. Also see *Ethnic Dolls*. Illustration **83**.

83. Oriental dolls, carved wooden heads and hands made in China; dolls originally dressed in Door of Hope Mission, Shanghai. Child dolls' clothes partly original.

ORIGINAL CLOTHES: Clothes on a doll when it was first sold would be called "original commercial clothes." But many dolls were sold without any clothes and the clothes were purchased separately or made at home. Therefore the clothes belonging to a doll during the childhood of the original owner can all be called original clothes. These original clothes are most important because they are the best and most precise key to the date of the doll and to the age and sex represented. Often it is difficult to differentiate between original clothes and *Contemporary Clothes*. Illustration **84**.

84. Original clothes on a doll having a wax-over composition head, eyes open and close with a wire at the waistline. Patriotic costume of the 1860s. (See index for other dolls in original clothes.)

ORIGINAL DOLL MADE IN A MOLD: A doll is produced by an artist creating an original model (usually in wax or clay) from which a master mold is created, the pressing or pouring molds are then made from the master mold and the original dolls are made from these pressing or pouring molds. The model created by the artist is not a doll since its function is purely a part in the process of manufacture and not as a child's toy.

PAPER DOLLS: Dolls made entirely of any type of paper; both two-dimensional and three-dimensional. They have changeable clothes made of paper and/or other materials such as cloth. The three-dimensional paper dolls do not necessarily have changeable clothes. See also *Two-Dimensional Wooden Dolls*. Illustrations **85A** and **85B**.

85A. Paper doll made by McLoughlin Brothers and named "Lilly Beers." It has paper dresses and hat.

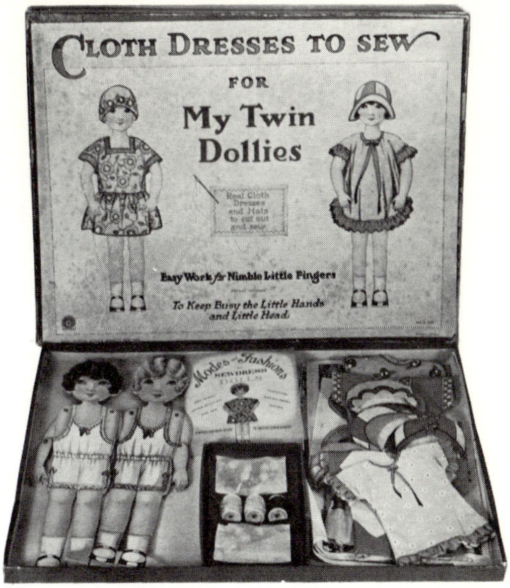

85B. Paper doll made by Samuel Gabriel Sons & Co. jointed at the shoulders and elbows: cloth dresses and hats.

PAPIER-MACHE: There are two principal forms of papier-mâché, the simple one
continued page 38

86. Papier-mâché head dolls, kid bodies, wooden limbs, early molded hair styles, probably made before 1860.

is made with laminated sheets of paper glued together. The other method which is nearly always used for dolls, comprises paper pulp and various other ingredients which form a type of composition. In dolls it is practically always impossible to distinguish between papier-mâché and composition. Papier-mâché is a form of composition, all papier-mâché dolls are composition, but not all *Composition Dolls* are papier-maché. Illustration **86.**

87B. Parian-type shoulder head of untinted bisque with floral decoration in the molded hair, glass eyes. The term "Dresden" was rejected for this type of decorated head.

PARIAN-TYPE DOLLS: Dolls made of fine white bisque (unglazed porcelain) without tinting. The features, hair and cheeks are painted (occasionally there are glass eyes). Eleanor St. George also used the terms "blonde bisque" and "Dresden" for these dolls. She also referred to early Jumeaus as being parian or blonde bisque. Her terms are rejected by the collectors. Illustrations **87A** and **87B.**

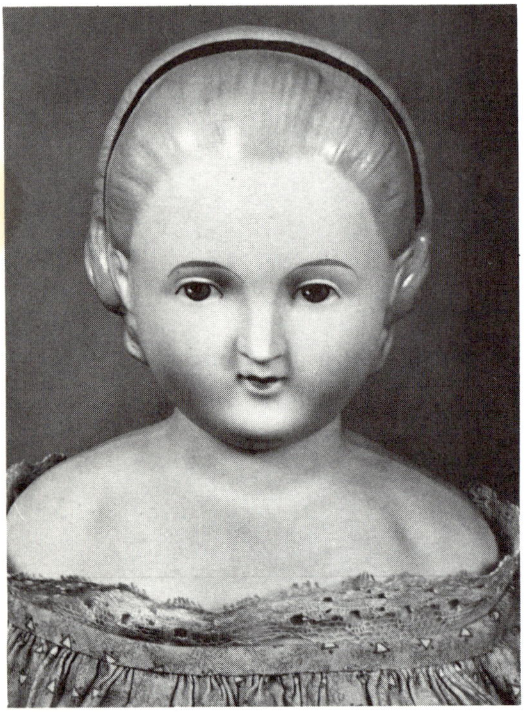

87A. Parian-type shoulder head of untinted bisque, molded hair with brush marks, panited eyes.

PARISIENNE: Emile Jumeau called his kid-bodied dolls "Parisiennes" and this term was also used by other makers for dolls associated with the Paris area. The majority of the collectors rejected this term. Illustration **88.**

88. Parisienne, rejected term for this type of lady doll with a bisque head and adult female shaped kid body.

PEASANT DOLLS: See *Regional Dolls*.

PEDDLERS: Dolls dressed to represent sales persons carrying their wares. Illustration **89**.

89. Peddler doll, composition head glass eyes, tray full of wares.

PEG-WOODEN DOLLS: Dolls made entirely of wood with mortise and tenon peg joints. Parts of these dolls were carved and parts were turned. The facial features and usually the hair were painted. They were made in the Grödner Tal, Austria from the end of the 18th century into the 20th century. The earliest ones have varnished heads while the late ones are merely painted and very crude. Often there is a carved yellow comb on the head of the early dolls. All during the 19th and early 20th century these dolls were known as *Dutch Dolls*, a term that was rejected by the collectors. Illustrations **90A** and **90B**.

90A. Peg-wooden doll probably made in the Grödner Tal area in the late 19th or early 20th century.

90B. Peg-wooden knee joint on the doll shown in **90A.**

PERIOD COSTUMES: Period costumes on dolls in *Original Clothes* representing an earlier period than that of the doll itself.

PINCUSHION FIGURE: Not chosen by collectors as a doll because it is used for purposes other than a child's plaything. This also applies to dolls' heads and upper torsos when used for other decorative purposes such as telephone dolls, lamp dolls, candy boxes, tea cosies, egg warmers, whisk brooms and so forth. Illustration **91**.

91. Pincushion figure of china, with molded regional costume. Mark of William Goebel.

92. Plaster doll of the chubby, googly-eyed type formerly distributed at carnivals, original clothes.

PLASTER DOLLS: Dolls made of a composition with plaster as the principal ingredient. Some of these dolls have a chalk-like appearance. Illustration **92**.

PLASTIC DOLLS: Dolls made of a soft plastic material such as vinyl, styrene and so forth. These dolls of soft plastic were produced after 1950. Also see *Hard Plastic* dolls.

PORCELAIN DOLLS: *Bisque*, *China* and *Parian-type Dolls*. These are all high-fired and translucent, but not transparent. In cases where the porcelain is thick it is almost impossible to see light through it. Illustration **93**.

93. Porcelain dolls from similar molds. Both dolls' heads have brush marks and pierced ears but there are slight differences in contours. Left, china head. Right, parian-type head.

PORTRAIT DOLLS: A doll having facial characteristics identifiable with a particular person. Illustration **94**.

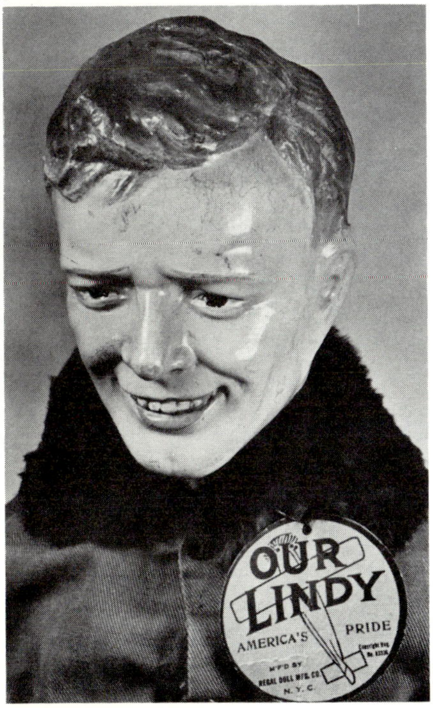

94. Portrait doll of Charles Lindbergh; composition head, hands and feet, cloth body; made by Regal Doll Mfg. Co.

PORTRAIT JUMEAU: An unacceptable term. It has been applied to any doll with an unmarked head on a marked Jumeau body. Dolls with various types of unmarked heads have been called a "portrait Jumeau." In order to be a *Portrait Doll*, a doll must have facial characteristics which are identifiable with a particular person.

POTTERY DOLLS: Low-fired ceramic dolls. These dolls are usually of a fairly thick ceramic. They break easily, are always opaque and can be glazed or unglazed.

POUPARD: A doll dressed in or delineated with swaddling clothes. Poupard is the *Trade Name* but collectors prefer to describe these dolls as *Swaddling-Clothed Dolls* (see also). The term poupard is sometimes erroneously applied to a *Marotte* (see also).

POURED WAX DOLLS: Wax is poured into a mold, when hardened around the edge the residual is poured off. The dolls are made partly or completely of wax. The wax may be beeswax, spermacetti (wax from a whale) or paraffin wax.

PRIMITIVE DOLLS: Dolls created by craftsmen having no formal or extended experience. These are generally somewhat crude and may or may not be sold commercially. Illustration **95**.

95. Primitive doll made of wood.

PROVENANCE: The provenance of a doll is the identification of its origin, its period and its type.

QUEEN ANNE TYPE DOLLS: Dolls made entirely of wood, except for those with eyes and limbs of other materials. The eyes when not painted are glass and occasionally of porcelain. The limbs are fabric or leather when not of wood. Dolls having the lower torso pointed, usually have leather or cloth limbs. The wigs are made of flax, hemp, or hair. The dolls from towards the end of the 17th century to about 1720 have carved faces, usually with painted eyes. Lord and Lady Clapham represent this type. Illustrations **96A**.

96A. Queen Anne type doll of wood, early period probably made between 1690 and 1720, original clothes.

Queen Anne type doll from about 1720 to 1780 have carved features usually with glass or porcelain eyes. Dolls from about 1780 to 1840 have gessoed features, the faces are somewhat flat except for the noses. The eyes can be painted or made of glass. Illustrations **96B** and **96C**.

96B. Queen Anne type doll of wood, middle period, probably made between 1720 and 1780, original clothes.

96C. Queen Anne type doll of wood, late period, probably made sometime between 1780 and 1840. Redressed.

REGIONAL COSTUME: Dolls in clothes representing the costumes worn in a particular region at a particular period. A region can be represented by several different outfits, namely everyday costume as well as a Sunday and/or festival costume. Many of these dolls were not made in the area represented and do not necessarily have the facial characteristics of the region represented. The clothes are usual-

continued on page 44

ly only rough approximations of the real costumes. Also see *Ethnic Dolls*. Illustration **97**.

97. Regional costume of Norway on a doll with a felt head and body.

REISSUE: Modern dolls made from old molds by the same factory or its successor, for example the modern china dolls' heads made by the Royal Copenhagen Porcelain factory from some of their 19th century molds. A reissue is not a *Reproduction* (see also).

RELIGIOUS AND CEREMONIAL FIGURES: Not classified as dolls. These include *Crèche Figures* (see also), Saints, Santons and so forth. Also excluded are the figures with the lower half in the shape of a round cage because these were generally used for religious purposes. Illustration **98**.

98. Religious and ceremonial figures include Santons, meaning "Little Saints." The Santons generally represent the common people of Provence, France, such as this couple.

REPLICA: Collectors overwhelmingly chose to use the word *Reproduction* (see also) instead of replica.

REPRODUCTION: Dolls made by producing a mold from an existing doll or figurine and using this mold for the reproduction. Illustration **99**.

99. Reproduction china head, made by Emma Clear.

RESTORED DOLL: A doll that has been repaired and/or repainted and/or with replaced parts. Cleaning a doll or restringing a composition body does not make a restored doll.

ROCK-A-BYE BABY: A *Trade Name* applied to several types of Baby dolls in the 1920s. This name should not be applied to the *Dream Baby* with an open mouth.

RUBBER DOLLS: Dolls made of either soft rubber or hard rubber. Rubber should noι be confused with *Gutta Percha.* Illustration **100**.

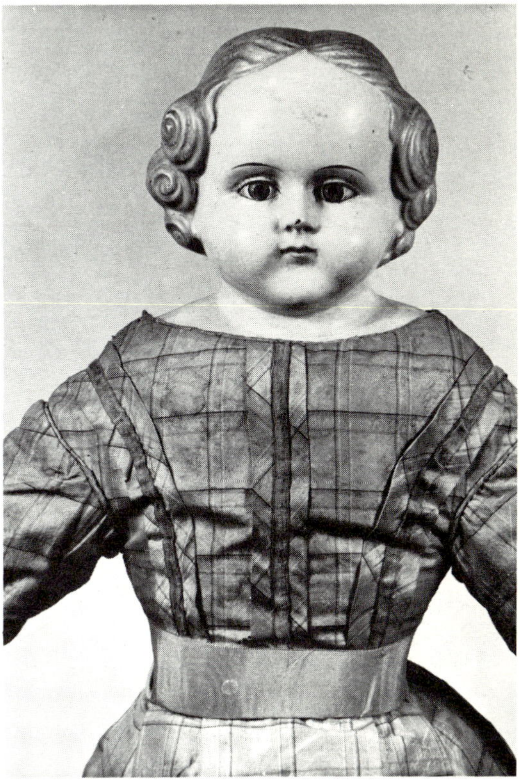

100. Rubber doll, head made of hard rubber by Goodyear.

SHOULDER HEAD: The head and shoulders made of the same type of material and placed on a torso of different material. The shoulder head may or may not have a neck joint. Illustration **101**.

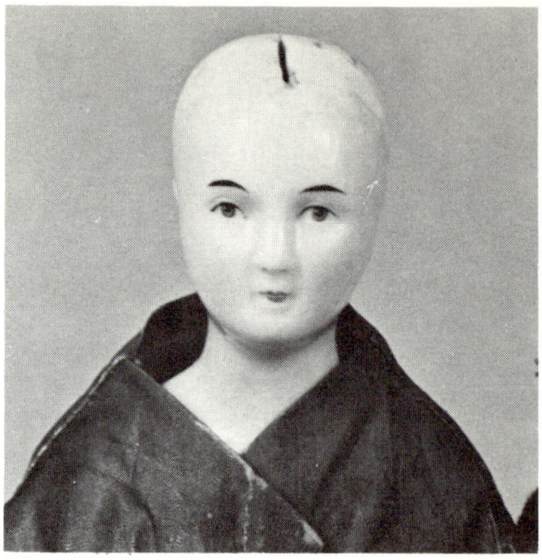

102A. Slit head of china for the insertion of hair, painted eyes.

101. Shoulder head of carved wood, probably made in Switzerland, in the 20th century.

102B. Slit head of composition for the insertion of hair, painted eyes.

SLIT HEAD: The slit from front to back at the top of the head with hair inserted. This method of attaching the hair was used primarily on wax-over composition heads, china and bisque heads. Illustrations **102A** and **102B**.

SHOULDER PLATE: The shoulder part of a *Shoulder Head*. This is usually separated from the head by a joint at the neck.

SNOW BABIES: All-bisque dolls or heads with caps covered with granular pieces of white porcelain. The all-bisque dolls also have the white granular porcelain on their bodies. See *Hatted Dolls*.
Illustration **103**.

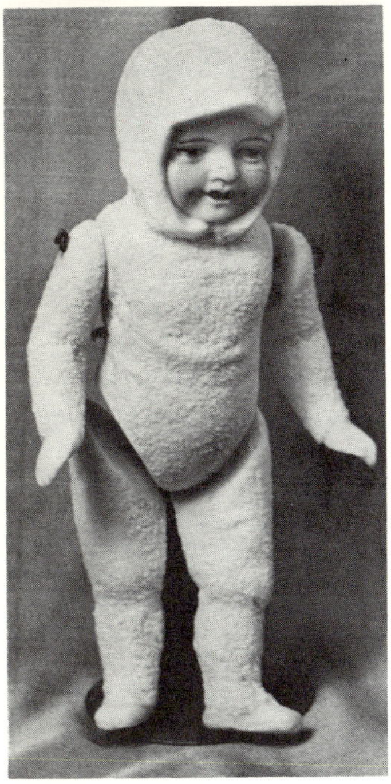

103. Snow Baby, all-bisque hatted doll covered with granular pieces of porcelain.

SOCKET HEAD: A doll's head having a neck joint wherein the hemispherical bottom of the neck fits into a socket in the top of the shoulders. A reverse socket head has the socket in the neck and the hemispherical projection on the shoulders. A socket head enables the doll's head to swivel in all directions.

SOLID DOME HEADS: See *Bald Heads*.

SOLID-WAX DOLLS: Wax poured into a mold and none of it poured off. The dolls are partly or completely made of wax. The wax may be beeswax, spermacetti (wax from a whale) or paraffin wax.

SPRINGFIELD WOODEN DOLLS: Dolls with wooden or composition heads, wooden bodies and metal or wooden hands and feet. The heads are molded and painted. These dolls were made in Springfield, Vermont in the 1870s and 1880s by the *Jointed Doll Company*, The Cooperative Doll Company and D.M. Smith & Company. Joel Ellis belonged to the Cooperative Doll Company. Martin, Sanders and Johnson belonged to the *Jointed Doll Company* and many of their dolls have a black band around the waist with the patent dates on it. *Mason and Taylor* belonged to D.M. Smith & Company. No proof has come to light as yet that many of the dolls referred to by collectors as "Mason and Taylor" dolls actually were made by them. Instead these were probably made by the *Jointed Doll Company*.

STAFFORDSHIRE DOLLS: See *China Dolls*.

STONE BISQUE DOLLS: Coarse bisque type dolls generally white. These dolls usually have a rough surface and the delineation is poor. The features are painted. This term applies to stone bisque head dolls as well as to all-stone bisque dolls. Illustration **104**.

104. Stone bisque dolls' head with a rough finish and poor painting.

SWADDLING-CLOTHED DOLLS: Baby dolls dressed in swaddling clothes. Generally the dolls have no feet and frequently they are also without arms. Often the clothes are painted and/or molded representations. Sometimes the bodies contain pebbles which can be rattled. The rejected *Trade Name* is *Poupard*.

TALKING DOLL: A doll that produces a sound by mechanical means, including bellows. This includes sounds made by squeezing, pulling or pushing a knob, tilting, winding or pulling a string. Sometimes the sound mechanism is activated by moving the doll's arm but this is not to be confused with the sound caused by friction or the movement of a spring strung joint.

TOPSY-TURVY DOLL: Doll having two heads with one concealed at a time. The term "double-end" and "turn-about" dolls were rejected by collectors. "Two-in-one" was a *Trade Name* for this type of doll. Illustration **105**.

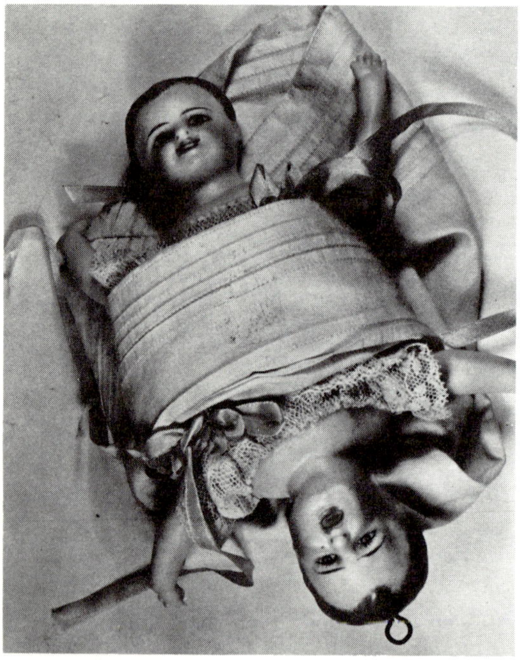

105. Topsy-Turvy doll with bisque heads and arms. One head is smiling, the other head is crying.

TRADE NAME: The name given to a doll or part of a doll by makers, assemblers or distributors and used in advertising and/or as a trademark.

TRIHEDRAL RUSSIAN DOLLS: All-wooden painted dolls made in Russia over a long period. These dolls are carved from a single piece of wood having three sides. They have no *Articulation*.

TWO-DIMENSIONAL WOODEN DOLLS: Dolls made of a flat piece of wood with or without *Articulation*. These are included as dolls.

UNBREAKABLE-TYPE COMPOSITION DOLLS: See *Cold-Press Composition*.

UNCUT CLOTH DOLLS: Back and front of the doll (or dolls) printed on cloth and designed to be cut out and sewed together. Sometimes the clothes are also printed on the cloth. Illustration **106**.

106. Uncut cloth doll named "Willie Winkie," made by Horsman.

WARDROBE OF A DOLL: Clothes and accessories that are additional to the ones on the doll. In order to qualify as a ward-
continued on page 50

107A. Wardrobe of a fashion-type lady doll with bisque head and lower limbs. The body is covered with twill. Clothes of the 1870s.

robe there must be at least two principal garments, besides the clothes worn by the dressed doll. A principal garment might be a dress, or a coat, or a cape, or a nightgown. In the trade "Trousseau" appears to be a synonym for Wardrobe. *Layette* is the term used for extra baby clothes. Illustrations **107A** and **107B**.

107B. Wardrobe of a child doll. The felt doll has clothing of felt and organdy. Trunk is marked "Cass Toys Athol Mass. Toys of Quality Made in U S A." Clothes are around 1930.

WAX-OVER DOLLS: Dolls with head or head and limbs dipped in wax. The wax can be over composition, wood, bisque or other material. Wax-over was strongly preferred to waxed by collectors.

WAX WITH INSERTED HAIR: The hair was inserted into the wax scalp in small groups or in single strands. This method was used in the 18th, 19th and 20th centuries. Illustration **108**.

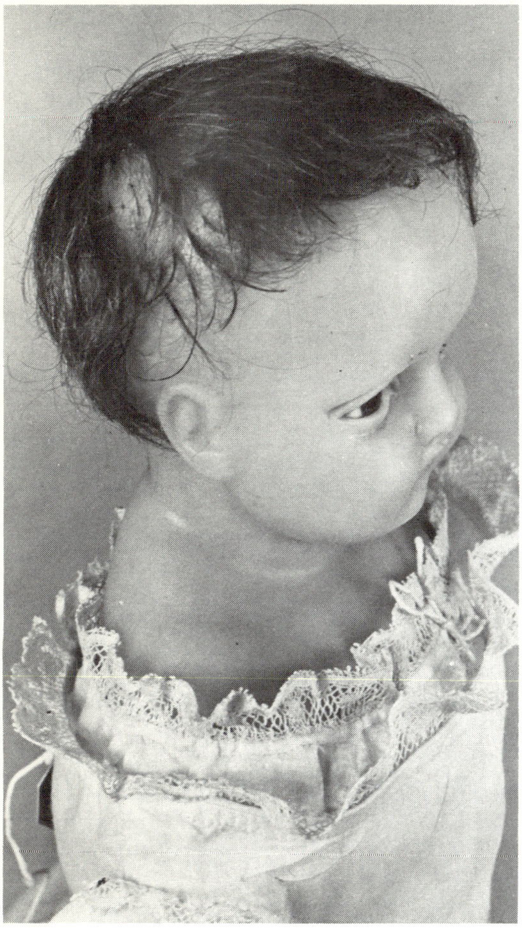

108. Wax with inserted hair. Poured wax Montanani-type dolls' head with hair inserted in groups.

WOOD-PULP COMPOSITION: The principal ingredient of the composition is wood-pulp or wood flour. This was used chiefly in making dolls after World War I by the *Hot Press* method (see also).

YARN DOLLS: Dolls made wholly or partly of yarn are not considered to be *Cloth Dolls.* This includes knitted and crocheted dolls of yarn as well as the dolls wound out of yarn.

INDEX

Q

R

S

T

U

V

W

Y